T0076206

AROUND THE WORLD IN EIGHTY GAMES

ALSO BY MARCUS DU SAUTOY

Thinking Better: The Art of the Shortcut in Math and Life

I Is a Strange Loop

How to Count to Infinity

The Creativity Code: How AI Is Learning to Write, Paint and Think

The Great Unknown: Seven Journeys to the Frontiers of Science

The Number Mysteries: A Mathematical Odyssey Through Everyday Life

Symmetry: A Journey into the Patterns of Nature

The Music of the Primes: Searching to Solve the Greatest Mystery in Mathematics

AROUND THE WORLD IN EIGHTY GAMES

FROM TAROT TO TIC-TAC-TOE, CATAN TO SNAKES AND LADDERS, A MATHEMATICIAN UNLOCKS THE SECRETS OF THE WORLD'S GREATEST GAMES

MARCUS DU SAUTOY

BASIC BOOKS
NEW YORK

Basic Books
Hachette Book Group
1290 Avenue of the Americas, New York, NY 10104
www.basicbooks.com

Printed in the United States of America

First Edition: November 2023

Published by Basic Books, an imprint of Perseus Books, LLC, a subsidiary of Hachette Book Group, Inc. The Basic Books name and logo is a trademark of the Hachette Book Group.

Print book interior design by name.

Library of Congress Cataloging-in-Publication Data
Names: Du Sautoy, Marcus, author.
Title: Around the world in eighty games : from tarot to tic-tac-toe, catan to chutes and ladders, a mathematician unlocks the secrets of the world's greatest games / Marcus du Sautoy.
Description: First edition. | New York : Basic Books, 2023. | Includes bibliographical references and index.
Identifiers: LCCN 2023006341 | ISBN 9781541601284 (hardcover) | ISBN 9781541601291 (ebook)
Subjects: LCSH: Games—Mathematics. | Games—Psychological aspects. | Mathematical analysis. | Problem solving—Mathematical models.
Classification: LCC GV1201 .D77 2023 | DDC 790.02/1—dc23/eng/20230523
LC record available at https://lccn.loc.gov/2023006341

ISBNs: 9781541601284 (hardcover), 9781541601291 (ebook)

LSC-C

Printing 1, 2023

TO MY SISTER CHARLOTTE

A WORTHY ADVERSARY ACROSS THE
GAME BOARDS OF MY CHILDHOOD

CONTENTS

CONTENTS

CHAPTER 0

OPENING MOVES

SIX TO START. Knight to f3. Two no trumps. Black stone on komoku. Take a chance card. Climb the ladder to square 38. These are the opening lines of some of the most wonderful stories that humanity has created. Since ancient times, civilizations have been playing games—some with dice, some with cards, others with pieces that move about on a board or tabletop.

Some have argued that our species should be called *homo ludens* rather than *homo sapiens* because it is the ability to play, not think, that has been crucial in our development. Evolution seems to have gifted humanity a penchant for trying out strategies and exploring imaginary worlds in the safety of a game, which allows us to prepare ourselves for real encounters. Among some of the earliest civilizations, sophisticated games emerged alongside the burgeoning of societies. Play is free. It is not real. It has its own location and duration. It is distinct from our ordinary lives. And yet it impacts on the way we behave outside the game: our wars, politics, arts, and sciences.

Like the stories shared around the campfire to bond our kinfolk, the games we have created allow us to share an exciting journey together in safety. Games resemble stories by conjuring up a fictional world, transporting the players into a different temporal dimension whose artificial barriers we enjoy striving to overcome.

Yet games are also stories that we can play out with our friends. As shared experiences, they have some advantages over the solitary disappearance into a book. They are more active, giving the players agency in the way the story evolves. A novel can move you to tears, but a game can make you feel guilty for your actions.

This book tells the story of the many crazy, fantastic, addictive games that our species has created during our time on this planet. To be sure, just as we do, other animals play games as children: lion cubs play fight before putting those skills to use as adult hunters. But our species is probably unique in continuing to play games into adulthood.

Not everyone grows up to become an adult game player. Some see the whole enterprise of battling to overcome self-imposed rules to reach an artificial destination as a complete waste of time. Playing a game is certainly an indulgence afforded to those with surplus time on their hands. But when we achieve the kind of utopia that absolves us of the need to work because the machines are doing it all for us, my belief is that games will become the central focus of our time thereafter.

I love games. I love them so much so that on all my travels around the world, I seek out the games that people like to play in the country I'm visiting. By the time I head home, my bag invariably has a few new local games packed inside. Just as the stories that people tell can help me understand a different society, I have found too that the local games people play can reveal much about the differences and similarities across different cultures. Some cultures favor games of chance over contests of strategy, perhaps reflecting their preference for a fatalistic outlook on life over a belief in agency over one's destiny. A fondness for a territorial game like Go over a more directly aggressive game like chess seems to reveal something about what a culture values and how it views the world. But while some games reveal the differences between cultures, other games seem so universal that versions are played around the world.

Games and their history help me to understand how each different culture has emerged. They are a living archaeology capturing the passions and pursuits of the people of the past. Tell me the game you play, and I'll tell you who you are.

And when I am stuck at home unable to travel, the games I've collected provide a way for me to escape back into the world they came from. The games are portals to the many countries I've visited. They are passports to other worlds—whether historical, geographical, or mathematical. Being cocooned by my collection at home is a constant and welcome reminder of the distances that we can all travel with our minds.

Yet I am not a historian, anthropologist, or psychologist, but a mathematician, and I think one of the reasons that games so resonate with how I see the world is that a good game shares much in common with the great mathematics that I have fallen in love over the years. The rules of the game are like the axioms of mathematics. Playing a game is like exploring the consequences of those axioms. Games for me are a way of playing mathematics. Time and again a game works because it embodies an abstract mathematical idea. A matching game like Dobble is successful because it exploits strange geometries in high-dimensional space. A board game like Ticket to Ride works because of the scoring system that balances risk versus reward.

The best games are those with simple rules that give rise to complex, rich, and varied outcomes. In the ancient game of Go, black and white stones are placed alternately on a 19 × 19 grid. The setup is simple, and yet the variety and complexity of the games that have emerged are extraordinary. For me this has much in common with the most beautiful mathematics. Prime numbers are simple to define: they are the indivisible numbers. Yet we have been playing with these numbers for two thousand years, and they continue to surprise us.

But there is another reason that I believe a mathematician is an ideal tour guide to the games humans play: because having a few mathematical tricks up your sleeve can often give the player an edge. Because games are defined by a set of rules that constrain the way you play the game, mathematics is a very natural language to explore the logical implications of these rules and to find the optimal path to get to your destination.

Ludwig Wittgenstein believed that it was impossible to define what a game is. *Game* was his principal example of a word that could only be understood through the act of using it: that is a game; this isn't. It was a process he called the *language game.*

This book is a game too. It can be regarded as an attempt to play Wittgenstein's language game. As I share the games that I have gathered on my journey around the globe, each will represent a new move in Wittgenstein's game.

My list of games conspicuously neglects to include sport. There is one exception, the Mayan ball game Pitz, which I couldn't resist sneaking in, but sports like soccer, basketball, and football have no place in the pages that follow. I'm not denying that football is a game, and one I love, but there is a reason that we came up with an additional word, *sport,* to classify certain types of games. For me, sport is a game of the body, while the games I am interested in are principally games of the mind. Probably my rather puny frame meant that as a kid I migrated to the nerds playing cards rather than the cool kids playing football. Individual preferences like mine do not lessen the importance of sport to humanity, since it plays a similar role in allowing a space to practice for danger. But as a guide and a mathematician, I have far more acute insights into mental games than into sports that hinge on brute strength and stamina. Some of the games I've included verge on puzzles, another concept in need of definition. In my book, a puzzle often gets elevated to a game once other people got involved. Wordle jumped from a puzzle to a game on a massive scale as soon as it added the feature to share your statistics with others. The element of competition or collaboration with others is what makes the difference. What, then, is an activity like solitaire? Is it a puzzle or a game, when to win the game, you must compete against the pack of cards? And what about computer games, pitting the player against the computer code. Wittgenstein is right: *game* is a very slippery concept to pin down. By taking you on a journey to experience my eighty games around the world, I am weaving together a thread, fiber by fiber, to hold together a coherent understanding of what we really mean when we say *game.*

There is another way in which this book is like a game: there are multiple ways to play. You can read the book from the first page to the last in the order that I have curated for you, proceeding through each of the numbered games we will encounter in sequential order. I have another suggestion, however: try throwing a six-sided die and beginning with the game that the die indicates. Once you've read that game, throw again. Add the number the die indicates to the number of the game you've just read, and proceed forward through the text as if it were a game board. Once you reach the end, simply loop back to the beginning, using the table at the back of the book to keep track of the games you've read on your journey. This allows each person to generate a different narrative journey by playing the game of the book, in turn creating a new book. (The mathematician in me couldn't resist calculating how many different books we can generate using this game. Working it out for yourself is one more game for you to play.)

I am not the first to suggest multiple ways to read a single book. B. S. Johnson's novel *The Unfortunates* consists of unbound pages jumbled in a box. The first and last chapter are fixed, but readers can choose the order of the twenty-five chapters that form the body of the book. That adds up to 15,511,210,043,330,985,984,000,000 different potential books. The French poet Raymond Queneau, cofounder of the experimental Oulipo movement, provided readers with even more options, composing a sequence of sonnets that offered readers a choice of ten different versions of each sonnet's fourteen lines.

This book is a celebration of the mathematics that swims seductively just below the surface of many of the games I love. Parallel to the main narrative, I offer passages that discuss these mathematical marvels. I invite you to dive into these fantastic mathematical stories that offer the keys to creating and often winning games. But if these mathematical dives leave you feeling out of your depth, feel free to return to the surface. Save them for your next journey through the book.

Please bear in mind as you read the book that it does not intend to be an encyclopedia of games, an exhaustive dictionary, or a compendium of rules. I have listed several comprehensive websites at the end of the book that can tell you how to play nearly every game imaginable. To those who will surely protest, "How come he didn't include that game?!" I must gently insist that this is my own very personal collection. I have selected those games that allow me to talk about the philosophy, culture, mathematics, and sociology of the games we play.

Each chapter corresponds to a different land and an exploration of the different games that the people play there. From India to China, Africa to America, we will play our way around the globe. These geographical chapters are separated by what I have called the book's "sea interludes." As I cross from one territory to another, I will use these liminal spaces as a place to explore some of the more philosophical, mathematical, or cultural themes around games that don't have such an obvious physical location.

When I compiled these games and game-related themes, it was a strange and unexpected miracle that the final count came to exactly eighty. Since this was as much a geographical journey through the world's games as a historical, cultural, or mathematical one, I couldn't resist using Jules Verne's famous story about Phileas Fogg's attempt to circumnavigate the globe in eighty days as a narrative structure for my story. Much like Fogg, I have been fortunate to use these games to venture on my own journey around the world, jumping between continents, making passages across sea and land. And mathematics will be an invaluable companion, my passe-partout, as I embark on my journey around the world in eighty games.

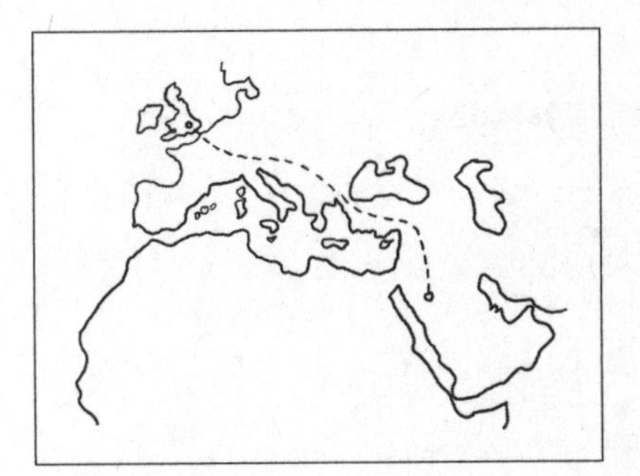

CHAPTER 1

THE MIDDLE EAST

WHAT MAKES A GREAT GAME? First, it shouldn't finish before it's even started. Even if you're not as good as your opponent, there should still be a chance that you can win. Second, it mustn't finish before it ends. The best games are those in which, right up to the last move, there remains a chance that anyone could win. This means that the best games invariably include an element of chance. And yet, third, chance alone is not enough. There should be some role for strategy and agency, or else the player is converted into little more than a machine implementing the rules of the game. Fourth, simplicity should give rise to complexity. Simple rules allow you to get playing quickly, but the variety provided by multiplying possible outcomes makes a game worth returning to again and again. Fifth, a good story is paramount. No need for castles and goblins, but still, like a good piece of mathematics, a rewarding game traces a compelling narrative arc.

My first game on our journey around the world I believe ticks all these boxes. It is one of the most perfect games and also, it turns out, one of humanity's most ancient.

1: Backgammon

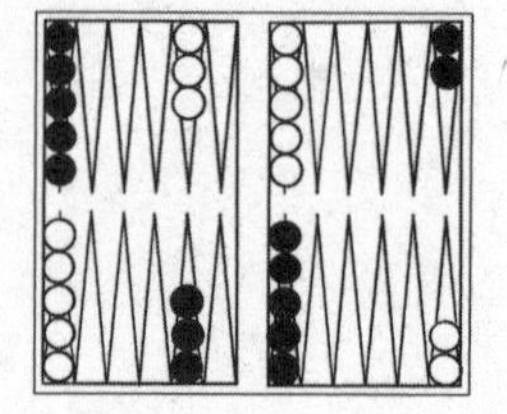

FIFTEEN BLACK CHECKERS. Fifteen white checkers. A board with twenty-four elongated triangles. The pieces race in opposite directions determined by the throw of two dice. Land on your opponent's lone checker, and that piece is sent back to the beginning of the race.

The game couldn't be simpler, and yet, in each of the thousands of matches I've played, the story is different every time. Backgammon is full of drama, twists, and turns. As in the mathematics of chaos, small changes can send the game in completely new directions. The lead can shift dramatically with a roll of the dice. A beginner has the chance to beat an expert, and even when the dice seem against you, strategic play can still give you the upper hand.

Backgammon is a game I came to quite late in life, while living in the Middle East. I was taught the game by a Bedouin in the desert who was able to communicate the rules to me in very minimal English. Within a matter of minutes, we were quickly gaming away with little effort, and I haven't stopped since. The richness of this game is in how simple it is to grasp, yet also how complex it is to play. It is also apt that I first learned this game in the Middle East because, as I will reveal, it evolved from the very first games played by the great civilizations of Babylon and Egypt.

It wasn't games but mathematics that first took me to the region. As a doctoral student I kept reading the papers of a mathematician who clearly was on the same wavelength as I was. We thought the same way. We were both obsessed by the same structures. It was like we were players of the same esoteric mathematical game, and once we knew about each other, it was clear we had to meet.

Although united by mathematics, it turned out that we were divided by politics. The mathematician was an Israeli settler on the West Bank. I'd been brought up on a diet of left-wing politics in Oxford fueled by the injustices of the Margaret Thatcher years. But the beauty of mathematics and games is that they provide a neutral universal territory in which opposites can meet. I traveled out to Israel to spend my first years after finishing my doctorate collaborating with this mathematical brother whom I had uncovered while reading the journals housed in our library in Oxford.

My time at the Hebrew University in Jerusalem was not confined to mathematics alone, and I loved traveling through the region to Jordan, Syria, and Egypt and especially staying with the Bedouin in Sinai. I was enjoying following in the footsteps of Lawrence of Arabia, whose portrait, in Bedouin dress, adorned the walls of my college in Oxford. And the most popular game that I found across the cafés of the Middle East was backgammon.

I've wiled away many an evening in the bare huts of the Bedouin in Sinai accompanied by a narghile (their name for a water pipe), a cup of tea, and a backgammon board. In Arabic the game is simply called *tawhah*, meaning "table," but the Israelis I've traveled with call it *shesh besh*. *Shesh* is Hebrew for "six," and *besh* is Turkish for "five." Throwing a six and a five is a great opening move, and traditionally you can let out an exited call of "shesh besh" to let everyone in the café know you got off to a cracking start.

In a region divided by politics, it's encouraging to see the popular game named after the union of two words from two different cultures. The power of games to unite warring factions might seem a strange idea given that games by their nature set up competition and rivalry. The huge preponderance of games of war across the world testifies to a strong linkage between games and real-life combat. And yet the board may be the site for competition and a battle of wits; the act of playing can still create community and connection.

Games for Peace, for example, is an initiative that taps into the power of playing a game together to break down barriers between

Israelis and Palestinians. Children from both communities begin by playing games online, working in groups to solve challenges and beat the other teams. The avatars they use mean that players are unaware of which ethnic group other players are affiliated with. Once the teams meet in person, kids invariably seek out the person behind the avatar they have been playing with only to discover that this new gaming friend is from across the political divide. It is a testament to the power of games to create trust and friendship out of competition.

Although Israelis know the game as *shesh besh*, I was always rather curious about the English name for backgammon, which sounded like it had something to do with choice cuts of ham. It turns out that the name refers to the idea of a back game, a strategy in which you leave pieces in your opponent's home section to mess up their play in the final stages of the game. It first became associated with the game around the seventeenth century, about the same time that doubles in the dice began to be rewarded with another doubling to speed up game play. A throw of two sixes means you can move four pieces, not just two, by six steps. This accelerated tempo is essential to the modern version of the game.

The name backgammon serves to describe two special ways of winning the game. If you manage to clear all your pieces before your opponent has even started removing theirs, then you receive a bonus for winning so conclusively. This is referred to in the modern game as winning a *gammon*. This doubles the money that you have bet on the game. What's more, if at the time of your victory your opponent still has pieces in your home quadrant waiting to find their way around the whole board, then you are said to have won a *backgammon*. This triples the stakes being wagered.

Whenever I play, I'm always pushed to remember the rather curious opening set up. It might be more reasonable if the game just started with all the pieces off the board. The theory is that the early part of the game, if played like this, is rather dull, and getting the fifteen pieces onto the board and interacting with one another takes some time. At some point in history, someone had the clever idea to start the game midway through. It's a bit like if two chess players got to know their openings so well that they could take a shortcut to the point where the game starts to take its own unique path. You could always decide to start a game from the point where things get truly interesting.

One of the distinctive features of backgammon is the way players move in opposite directions. This leads to exciting interactions throughout the game. Once the pieces have finally moved past each other, the game becomes simply a race to clear your pieces off the board. But the drama of backgammon often comes from the friction that ensues along the way, such as when you manage to land on your opponent's unprotected counters and send them back to the beginning again. But caution is in order. That captured piece can wreak havoc on your pieces as it makes its way back around the board.

What I love about backgammon is that it encapsulates many of the characteristics that make a great game: the rules are simple; yet outcomes are rich and varied. It's got drama and surprise—the dynamics of who's ahead can change on the throw of the dice—yet strategy is important in placing pieces to block your opponent.

Not only does backgammon embody some of the key characteristics that make a good game, but it also traces its origins to some of the first games played by our ancestors. These were not games of war but racing games, where the aim was to clear your pieces off the board before your opponent.

Although backgammon was a game I came across late in life, it evolved from a racing game that I encountered when I was first falling in love with games as a kid.

2: The Royal Game of Ur

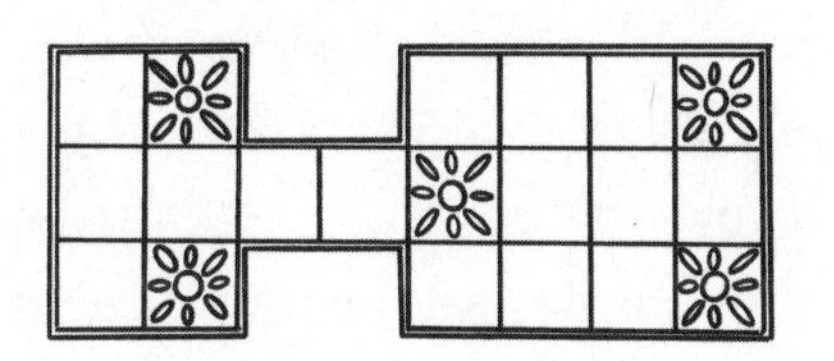

IT WAS IN THE REFORM CLUB on Pall Mall in London that Phileas Fogg made his bet that he could circumnavigate the globe in eighty days. Just twenty minutes' stroll up the road is the place that ignited my own quest to travel the world in search of games: the British Museum. When I was I child, my grandparents lived just around the corner, and I loved to drop by the museum, roaming the corridors and feasting on the extraordinary artifacts on display. I still remember the Saturday morning that I wandered into the Middle Eastern gallery and discovered the Royal Game of Ur.

The game was a thing of beauty. Its board was made not of cardboard, like most of the games I had at home, but of shells inlaid in a delicate wooden box colored in reds and blues. The game got its name from the Royal Cemetery of Ur, a site located in modern-day Iraq, where it was discovered during excavations by archaeologist Sir Leonard Wooley in the 1920s. The game that was unearthed is thought to date back to 2600 to 2400 BCE.

The board consisted of twenty squares made from shells carved with stunning symmetrical patterns. These squares were arranged in two rectangles of 3 × 4 and 3 × 2 squares with a narrow bridge of two squares joining them. Next to the board was a set of counters, to be placed by each player on the board, and the dice that would be used to play the game. Intriguingly the dice were not shaped like the cubes I was used to throwing on my Monopoly board at home but instead consisted of little triangular-based pyramids that mathematicians call *tetrahedrons. Tetra* is Greek for "four" and corresponds to the four triangular faces of the pyramids.

Using a pyramid as a die is a curious experience. When it lands after you throw it, there is no face showing, only a corner pointing up at you. Careful scrutiny revealed that two of the four corners of the pyramids were colored in white, and it seems that players were expected to throw a handful of pyramids and then count the number of white corners showing. The total told you how many squares you could move forward during that turn. Given that a tetrahedron has four corners and two are colored white, there is a fifty-fifty chance of it landing white spot up. Mathematically this has the same effect as tossing coins and counting the number of heads that are showing.

Still, questions abounded about this game. When I first discovered the board in the British Museum, no instruction manual had been found, and there was little to do other than speculate about how pieces might move along the board. It wasn't until years later, when I met museum curator Irving Finkel, that I learned that a set of instructions for the game had since been discovered. The great thing about the Babylonians is that they carved their texts into tablets using a script called *cuneiform*. A reed stylus would be used to make indentations in wet clay that would then be baked in the sun or a kiln to preserve the text. Unlike paper, the resulting tablets were able to survive for thousands of years. When one of these ancient tablets came to light seeming to refer to a game, Finkel took up the challenge of decoding the rules.

The tablet was written by Babylonian scribe Itti-Marduk-balatu on November 3, 177 (or 176) BCE, which we know because the scribe obligingly signed and dated his text. This meant that it was created a lot later than the game board on display in the museum. The tablet has a grid marked on one side consisting of vertical, horizontal, and diagonal lines and two columns of closely written text on the other. Finkel was also able to make use of photographs of an earlier tablet that had been destroyed during World War I to prove that this earlier tablet was describing the same game. This earlier tablet also revealed the true name of the game: the Pack of Dogs.

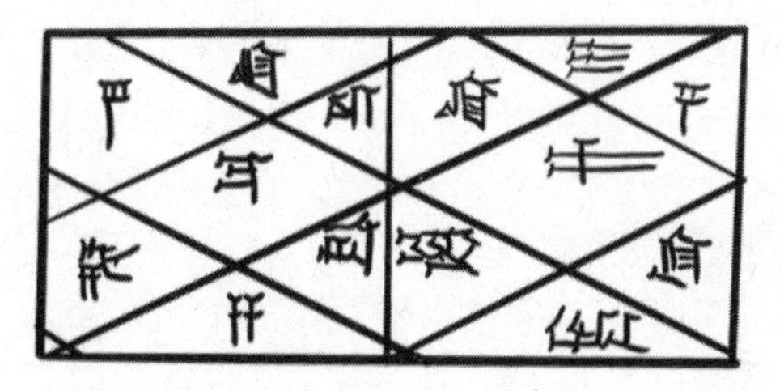

At first sight the grid pictured on the tablet looks nothing like the board that the game is played on. The grid portrays eighty-four cells marked out by crisscrossing lines. Still, this turns out to be an essential key to understanding the game. Finkel understood that the drawing should be read as showing twelve sets of seven cells with a central cell surrounded by six others. Inscribed in these central cells were the twelve signs of the zodiac. The temporal order of the signs of the zodiac therefore translated into an order in which these cells in the grid should be read. The six cells around the sign of the zodiac corresponded with a string of strange sentences: Scorpio: "You will draw fine beer"; Gemini: "You will find a friend."

This game, it seems, doubled as a fortune teller. Throughout the early history of games, it turns out, there was a close bond between games as leisure activities and means of predicting the future. For the ancients, the way games mimicked life seemed almost magical. The randomness of the throw of a die was a perfect metaphor for the way that one's fate is determined by external forces.

The game was clearly popular—hundreds of similar boards spanning a period of two thousand years have been discovered in Iraq, Iran, Israel, Jordan, Lebanon, Turkey, Cyprus, Egypt, and Crete. There is even evidence of a very similar game being played in the early twentieth century by the Jewish community in Cochin in India.

Some of these other boards have a different layout, which actually helps make sense of the direction of play and tells us which squares correspond to signs of the zodiac. Rather than the bridge of two squares running into the six squares arranged in a 2 × 3 configuration, these boards featured a long line of eight squares running away from the twelve squares in the 3 × 4 configuration. This meant that the central line of squares was itself twelve squares. It is these squares that seem to correspond to the twelve signs of the zodiac.

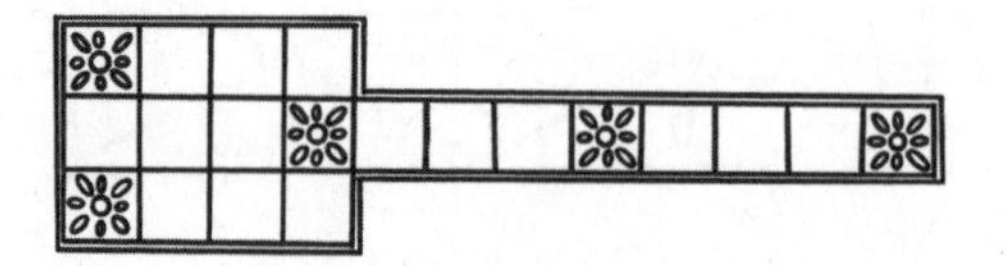

The two columns of cuneiform on the reverse side of the tablet give us the rules of the game in detail. They represent the oldest-known rules for a board game, and yet, even after being translated, their instructions are far from transparent. It is possible that the scribe was actually recording an advanced version of the game played by people who already knew the basic version.

The instructions help us to understand that the players each have five pieces named after birds: a swallow, a storm bird, a raven, a rooster, and an eagle. The pieces look identical, so it isn't clear if they genuinely had different roles to play according to the birds they represented. (It is believed that chess was actually the first game in which pieces were assigned different functions in game play.) Still, it is possible that, for example, the birds were intended to enter at different points on the board. The game was called Pack of Dogs because the word *dog* is used in many ancient languages to refer to a gaming piece. One might ask, then, why choose names of birds for the pieces that are otherwise referred to as dogs? One interesting explanation suggests that this has to do with birds passing through the twelve signs of the zodiac, potentially serving to represent the five visible planets as they pass through the twelve constellations of the zodiac. Mars, for example, has traditionally had some association with the raven—the rest of the birds may have had similar celestial counterparts.

The hunch is that players were supposed to sit on either side of the board and play their pieces along the four squares on their side of the board before turning onto the common racetrack of twelve squares shared by both players. The aim was to be the first to get your pieces to the end of the track before your opponent could land on your piece, sending it back to the beginning of the board.

The beautiful symmetrical designs inscribed in the board in shell are meant to do more than decorate the spaces. Rosettes marked special squares where pieces were safe from capture. Also, according to the rules of the advanced game detailed on the tablet, landing on these squares earned your piece a reward. The tablet describes these rewards in rather colorful terms: if the swallow lands on a rosette, it is rewarded with "success with a woman." The other birds earned food or beer. These translated into a different points tally corresponding to the probability that each bird might land on a rosette. The eagle, for example, only enters on the tenth square and so only has two possible rosettes it can land on before it finishes the race. It therefore scores highly if it manages to land on one of these two rosettes. The cuneiform instructions imply that if a piece passes over a rosette without landing on it, then a penalty should be paid from your points tally. The tablet colorfully refers to this as "starvation" or "failure with a woman" for the offending piece.

The resulting game is remarkably sophisticated. It is not simply a racing game but one in which points are being scored, with different roles played by different pieces as they enter the game and an additional narrative layer provided by the possibility of fortune telling.

This game took a long and complex journey to arrive in the British Museum. My nine-year-old self didn't question why a museum in London should be home to a game from Mesopotamia. But returning to gaze on the game as an adult, I am acutely aware that the fascination of earlier generations with collecting artifacts from around the world robbed those cultures of their heritage. It is a valuable reminder as I make my own journey around the world: there is a fine line between celebrating a culture through an exploration of its games and plundering those games disrespectfully for my own amusement. Even if a physical game isn't stolen, an idea can be. My journey around the world would quite often reveal to me that games I thought were Western in origin had been inspired without attribution by games in India or Asia.

Its fraught path across the world notwithstanding, the Royal Game of Ur stands as a remarkable monument to the deep-seated human impulse to create and play games. Finkel's decoding of the rules has provided the game with a new lease on life. On my recent visit to the museum as an adult to retrace my childhood encounter with this ancient game, I discovered that the shop now sells replicas of the Game of Dogs or the Royal Game of Ur, as it is still known. I couldn't resist buying a copy to take home. It's not as beautiful as the original, and the instructions are printed on paper rather than carved into a piece of clay. But it is a remarkable thing to be able to play a round of the same game that entertained the Babylonians five thousand years ago.

3: Senet

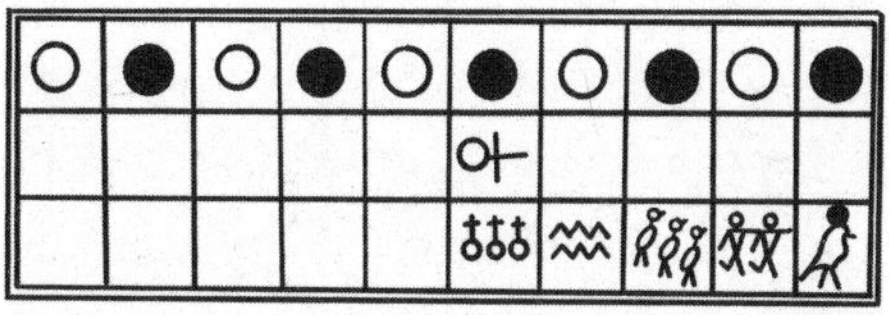

NOT TO BE OUTDONE by their Babylonian neighbors, the ancient Egyptians had their own racing game called Senet. Rather than using the strange-shaped board of the Babylonians, the Egyptians played on a simple 3 × 10 board of thirty squares. Paintings of people playing the game have been found on walls of tombs, one dating back as far as the twenty-seventh century BCE belonging to Hesyre, the chief dentist and an avid games collector. Papyrus doesn't last as long as a good Babylonian clay tablet, which means that no set of instructions has survived, but the images of people playing the game still give some hint as to its rules.

Players placed their five pieces alternating at the beginning of the board. Very often each player had different-shaped pieces, which might be described as pawns and spools, or lion heads and dogs. The game consisted of a race to the end. Pieces could be captured and sent back to the beginning, as in backgammon, but pieces were considered safe from attack and protected from being jumped if they were in groups of three.

I still remember the excitement in London when the treasures of Tutankhamun were exhibited in 1972. While most of my friends were obsessed with the mummies, the amazing gaming boards caught my attention—since Senet boards were not just documented in pictures but buried alongside other treasures bound for the afterlife. Of the five versions of the game in Tutankhamun's tomb, one mounted on a wooden sleigh-like structure was particularly impressive. Made of ebony and ivory, the board had a drawer underneath in which the game pieces would be stored. Players who couldn't afford such ornate boards made do by carving the 3 × 10 grid into the flat rooftops of buildings. One of the most recent examples of the board dating from the first century AD was found sketched on the roof of the hypostyle hall of the Temple of Hathor at Dendera in Egypt.

There is a special significance in the Egyptians' decision to bury this game alongside dead pharaohs. The name Senet means "passing through," because it was not just a game but also a narrative depicting the passage into the afterlife. At the end of each board, five squares were marked for the particular significance they carried both in game play and in the journey from life to death. The fifth square from the end boasted a circle connected to a "+" symbol and was referred to as the House of Happiness. Land here and you were safe from being captured or jumped by your opponent. The next square along, marked with an X or wavy lines, was called the House of Water. Land here and you would be sent back to a square in the middle of the board called the House of Rebirth. The last three squares had the numerals III, II, and I, each denoting the score needed to play the pieces off the board and into the hereafter. Some boards replace the numbers with birds representing the soul on its final journey. Senet was at once a game and a depiction of the quest for immortality in the afterlife.

In the possession of the British Museum there is a beautiful papyrus showing Hunefer, a royal scribe, playing the game of Senet in the Book of the Dead that describes his own burial. It seems probable that he might even have written his own Book of the Dead given that he worked as a scribe. On a page illustrating the ceremonies carried out at the entrance of the tomb on the day of burial, Hunefer is seen sitting in a booth playing the game as if already fighting fate for his place in the afterworld.

Another depiction of the game, dating from 2300 BCE and found on a wall in the tomb of Pepi-Ankh at Meir, includes some fantastic hieroglyphic banter between the players. One player boasts, "It has alighted. Be happy my heart, for I shall cause you to see it taken away." But his opponent isn't ready to concede quite so quickly: "You speak as one weak of tongue, for passing is mine."

As a mathematician, I regard the Babylonian and Egyptian civilizations as representing a turning point in our species' development. They were the first to move from simply counting and measuring to doing really interesting mathematics. It's striking that these two civilizations were also the first to create intricate rule-based games, which suggests that there's something to my hunch that mathematics and games have a deep and intimate relationship.

To create a game is to set out rules for how pieces on the board behave. Those rules should be simple, but the logical consequences should be rich and varied. They should be consistent and make sense, not resulting in confusion or contradictions. In creating these games, both civilizations demonstrated a recognition that the world around them was bound by mathematical rules. They also revealed a confidence that it was possible to understand those rules and even exploit them to tame your environment. Understand the patterns behind the flooding of the Nile and you could win the agricultural game of life. Master the mathematical rules of construction and you could build an empire as great as Babylon's.

Yet the fun of playing games was still in a primitive state. While Senet shares much in common with the racing style of the Royal Game of Ur, it is far more tedious, and it certainly doesn't match the drama of backgammon. Your piece gets stuck at the end of the journey, waiting to make its move to the end of the board. Senet just doesn't have the same ebb and flow as the Babylonian game. Moreover, although both civilizations were great at mathematics, I would also rate the Babylonians as the more mathematically accomplished. Take the way the two wrote their numbers. The Egyptians had to invent new hieroglyphs for ever larger numbers. The Babylonians had cottoned on to the clever idea of the place-number system.

That said, the Egyptians made great strides in geometry, coming up with a clever formula for the volume of a pyramid (useful for working out how many blocks would be needed to build the amazing structures in the desert of Giza). But the Babylonians had already found algorithms for solving quadratic equations, useful for—well not really useful for much but a fun exercise anyway, since they had taken up the idea that mathematics might be a source of enjoyment. Both civilizations realized the importance of pi and came up with different strategies to calculate the ratio of the circumference of a circle to its diameter. The Egyptians estimated it to be about $\frac{256}{81}$ (roughly 3.16), while the Babylonians estimated pi to be about $\frac{25}{8}$ (3.125). As we shall see, playing a game might actually have led to some of these accurate estimates for this important number.

The British Museum has a rather hilarious papyrus dating from 1250 to 1150 BCE depicting animals playing board games, a theme repeated many times through history.

Many have celebrated both the Royal Game of Ur and Senet as early forerunners of backgammon. Both games are about racing, not fighting. The Royal Game of Ur has the aspect of different directions of play for different players that we find in backgammon. Both games see counters returned to the beginning of the race if they are captured. Due to its better game mechanics, the Royal Game of Ur is successful outside Babylon, while Senet boards are confined to the Egyptian Empire, though they managed to survive for thousands of years even so. Perhaps the fact that it was seen not just as a game but as an integral part of the process of death and the passage to the afterlife gave it such lasting resonance in Egypt. Other ancient artifacts testify to a similar deep-seated reverence for death, fate, and chance.

4: Rolling Bones

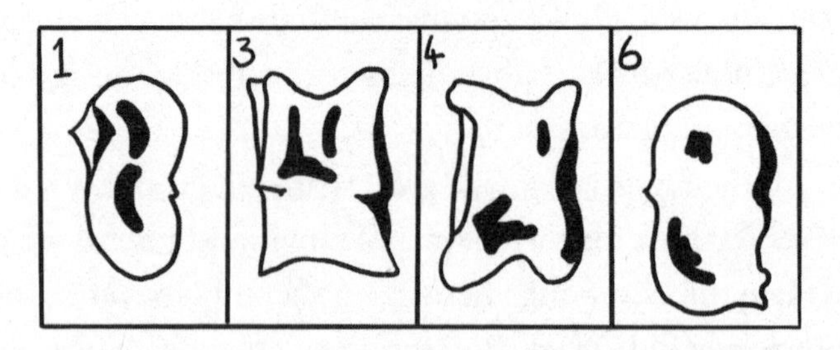

THE ONE INVENTION THAT backgammon, Senet, and the Royal Game of Ur all rely on is the concept of the die. This object of chance is probably more ancient than all these games. I discovered this when I played the role of amateur archeologist for a day.

Some years back, on a trip to Israel, I took my children to an archeological dig in Beit Guvrin. I was frankly rather surprised that they let a family of untrained archeologists loose on such an ancient site, but the leader of the dig explained to us that it was because the site was such a popular location for people to settle in ancient times. Consisting of layer upon layer of ancient settlement, there simply was so much stuff in the site that the archeologists were happy to tolerate a few broken pots in exchange for our unskilled labor.

We did indeed dig up a good range of bits of pots, some with interesting designs though most just boring bits of clay. We also dug out a good number of bones that we assumed were the remains of ancient dinners. But the archeologist guiding our dig surprised us by announcing that these were in fact one of the earliest examples of dice, known as knucklebones. Such bones are predominantly the ankle bones of hooved animals like sheep, bones that are called *astragali* in the trade. It is generally believed that, unless animal ankles were some culinary delicacy at the time, these bones were chosen because of their particular shape.

Ankle bones are almost symmetrical, and like little extended cuboids, they naturally fall on one of four sides when thrown, making them very good candidates for use as dice. The preponderance of these particular animal bones in sites like that at Beit Guvrin suggest that ancient humans were already enjoying games of chance. The archeological site in Beit Guvrin only dates back some two thousand years, but other examples have been found in sites that are seven thousand years old.

In August 2022, a rather extraordinary collection of bones was dug up in Beit Guvrin. Most of the knucklebones that had been dug up in the site previously didn't have any inscriptions, but this new collection of 530 bones was rather different. Dating from the Hellenistic period twenty-three hundred years ago, some of the dice had names of Greek deities, like Aphrodite, the goddess of fertility, and Hermes, the god of trade, carved on their sides. Others featured more down-to-earth inscriptions, like "thief" or "prisoner" or "stop." Tablets found alongside the dice included instructions for how to play, and a few of the bones were filled with lead to even up the throw of the dice.

The reason the find is so significant is that most of the bones that had been excavated, including the bones we dug up, didn't have anything carved into the sides. In the words of Israeli archeologist Dr. Lee Pri-Gal, the find was "very rare in both quality and quantity of the inscriptions on the bones. The collection indicates that in times of doubt, humans were searching for answers through external forces, like magic and gods. In ancient times, people—and women in particular—dealt with constant uncertainty and tried to defend themselves using magic. Since these dice were considered lucky charms, people would place them at the entrance to their home, hoping they would bring them luck."

It appears that the asymmetry of the bones might have been important in scoring different throws. The Romans had different names for each side according to their shape: the belly, the hole, the ear, and the vulture. In Mongolia, where the gaming bones are called *shagai*, different sides are named after animals: horse, camel, sheep, or goat. The bones can sometimes rarely land on a fifth side, although this is pretty unstable. The Mongolians refer to this side as the cow.

The different shapes of the faces translate to different chances of their landing face up. That in turn means that the rarer sides can score different points than the more common sides. The sides corresponding to the belly and the hole in Roman terminology are much broader than the other two sides. Roman gamers therefore assigned these rarer sides scores of six and one and the flatter sides scores of three and four, accounting for the fact that both extreme good and bad luck happens rarely. Modern analysis of these bones has revealed that the rare sides have a one-in-ten chance of being rolled, while the broader sides each land about four out of ten times.

In the cuneiform tablet describing the advanced rules for the Royal Game of Ur, the scribe actually refers to the use of two different-sized ankle bones: one of a sheep and one of an ox. Finkel speculates that having thrown the sheep die, players could gamble on doubling up the score or losing their go by throwing the ox die, rather similar to the idea of the doubling cube in backgammon, which I will come to at the end of the chapter.

The Romans are known to have played games casting four of these bones, although Roman law required that gaming be limited to Saturnalia, the Roman holiday in December in honor of Saturn. Throwing a complete set of ones was sometimes known as "scoring a dog"; sometimes the roll was referred to as "the vulture." It seems that throwing "Venus," an outcome where all sides are different, was regarded as a winning move. As the emperor Augustus writes, "We gambled like old men during the meal both yesterday and today, for when the dice were thrown whoever turned up the dog or the six put a denarius in the pool for each one of the dice, and the whole was taken by anyone who threw the venus." The emperor Claudius was so addicted to gaming that Suetonius documents in his *The Lives of the Emperors*, "He actually used to play while driving, having the board fitted to his carriage in such a way as to prevent his game from being disturbed."

The bones not only were used as dice but also appeared in versions of games still played worldwide today. Players could demonstrate manual dexterity by picking up the bones in turn in some prescribed manner. A player must throw a ball in the air with one hand and pick up another bone with their other hand before the ball lands.

Once again these games also doubled up as a way of telling fortunes. Romans would enter the temple, and either they would pick up four or five dice themselves or the priest would cast them as a way of asking the oracle to predict their future. A result of three fours and two sixes, known as "the throw of child-eating Cronos," was accompanied by the following omen: "Abide in thy house, nor go elsewhere, lest a ravening and destroying beast come nigh thee. For I see not that this business is safe. But bide thy time."

Tibetan temples also used dice and an 8 × 7 board in order to predict what a person might be reborn as after reincarnation. However, as

Laurence Waddell, a Victorian scholar of Buddhism, discovered on a visit to Tibet, the priests were not exactly playing fairly. "The dice accompanying my board seems to have been loaded so as to show up the letter Y, which gives a ghostly existence, and thus necessitates the performance of many expensive rites to counteract so undesirable a fate." The priests were using loaded dice, then cashing in on the bad omens that resulted.

As time went on, it seems that games demanded a shape where no side was favored over any other. Symmetry was the solution. And so players started to carve these bones into shapes that would make a fairer kind of dice. As an unexpected side effect, this exploration may well have led to the mathematical discovery of the Platonic solids.

5: Symmetrical Dice

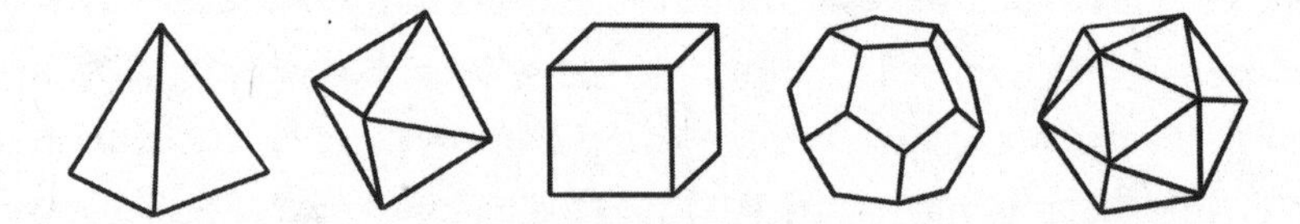

GIVEN THAT THE KNUCKLEBONES landed on one of four sides, it is perhaps understandable that the first purely symmetrical dice to appear on the scene took the form of the little tetrahedrons that so perplexed my younger self when I first glimpsed the Royal Game of Ur. The first gamer seeking to carve a set of knucklebones into a four-faced shape that was symmetrical would have produced four faces in the shape of equilateral triangles, which would land with a point sticking up in the air. Though the evolution of a knucklebone's elongated cuboid into a symmetrical tetrahedron was a significant step forward, a few steps remained before humans would hit upon the six-faced cube that most of us are used to gaming with today.

Cube-shaped dice made from clay or pottery started to emerge at the beginning of the third millennium BCE. At a site in northern Iraq called Tepe Gawra, archeologists have discovered recognizable dice with pips marked on the sides from one to six. Similar terra-cotta dice have popped up in northern India and Pakistan from the Harappan civilizations created between 2600 and 1900 BCE. These ancient dice are laid out with a two on the face opposite the face with a three, a four opposite a five, and a six opposite the one. It wasn't until cube-shaped dice emerged in Egypt in 1370 BCE that the modern arrangement arose, where opposite sides add up to seven. Besides suffering from an uneven shape, knucklebones lack uniform density, which meant that even symmetrical dice carved from bone could be imbalanced by pockets of air. Gamers must have discovered that repositioning numbers so that a high score on one face was paired with a low score on the opposite face served to even up any asymmetry that the cube might have.

Various texts reveal that dice played a part in many different cultures. The Hindu epic the *Mahabharata*, written around 300 BCE, depicts a game of dice between Yudhisthira and the Kauravas. The Kauravas owned a pair of dice that magically did their bidding and produced whatever numbers they desired—in other words, the dice were rigged. Yet Yudhisthira was so convinced that you couldn't predict or control the outcome of the dice that he lost all his wealth, his kingdom, his brothers, and eventually his own freedom, ending up in exile for thirteen years.

The Old Testament often mentions the casting of lots to make decisions, which some have interpreted as the throwing of some sort of dice. They feature, for example, in the story of Jonah and the whale. The sailors used dice to decide which passenger to throw into the sea to appease the storm that was battering their boat, singling out Jonah as the victim. His luck changed when he was swallowed by the whale, which spat him out alive and well three days later.

The cube and the tetrahedron are not the only symmetrical shapes available for gaming, as any player of Dungeons & Dragons knows well. The discovery of more esoteric shapes for dice came later and can be seen in far-flung sites around the world.

There is evidence that at the same time as the Royal Game of Ur was being played with its tetrahedral dice, another contemporary culture was starting to explore symmetrical shapes that would eventually become recognizable as dice. Hundreds of carved balls made of basalt or sandstone, dating back to 2500 BCE, have been discovered in the heartland of Aberdeenshire in northern Scotland. Etched into the sides of the balls are geometrical patterns. The Neolithic sculptors appear to have played around with different symmetrical arrangements of knobs, which almost resemble patches on a soccer ball.

The Ashmolean Museum in Oxford has a wonderful collection of these stone balls, which I recently had the chance to see in person. As big as a closed fist, the balls are very satisfying to hold in the hand. As well as arrangements of six patches and four patches corresponding to the symmetry of the cube and tetrahedron, the balls reveal that the sculptor was experimenting with other possibilities, including 8, 12, 14, and even 160 faces. Because my own research is dedicated to trying to discover new symmetrical objects in higher dimensions—the dice of hyperspace, you might say—there is something particularly moving about being allowed to hold the stones that humans were using to explore symmetry five thousand years ago.

It is not clear what purpose these balls served for the Neolithic culture that carved them. There is some speculation that they were used in divination. Some have suggested that they were used as symbols of authority by clan leaders. They have never been found in tombs, and so it has been suggested that they may have been significant more to the tribe than to a particular member of the group. They don't seem to be obviously part of any game, as far as we can tell.

But perhaps we are being too closed-minded in trying to find a definite use for these shapes. Perhaps this is evidence of the first mathematicians emerging and being fascinated by the possible symmetries they could make. These sculptors might have discovered that shapes with eight or twelve faces seemed easier to arrange than those with other numbers. They were onto something: two new possible die shapes beyond the cube and tetrahedron.

It isn't until 500 BCE, near Bologna, that these new shapes appeared on the scene: the octahedron, made up of eight equilateral triangles, and the dodecahedron, with twelve pentagonal faces. It was the

Romans who discovered that twelve pentagons could be carved out of a ball of stone in such a way that no face was favored over any other. The symmetry made it an ideal candidate for a new die. Examples of these dice have featured Etruscan-Roman numbers carved onto the twelve faces of the shape. Archeologists have also recovered examples of this shape carved with the twelve signs of the zodiac.

The twelve-pentagon die is quite a sophisticated shape. The Romans may have been drawn to carve out these twelve-sided symmetrical shapes because they were familiar with fool's gold, a compound otherwise known as pyrite, which often arranged itself into eye-catching crystals. As it was often found alongside copper, miners would have been used to seeing it both in its cubic form and in large lumps made up of pentagonal crystals. The crystals are not completely symmetrical and wouldn't be suitable as dice on their own. But their distinctive shape might have provided the inspiration for Roman sculptors to discover that they could actually cut the sides so that each was a perfect pentagon.

Up until this point in history, the making of dice seems the product of creativity and experimentation and not what we would regard as a linear scientific endeavor. The first appearance of the octahedron and dodecahedron in antiquity marks the arrival of the first real mathematicians on the scene. And these mathematicians were interested in a more analytic approach to dice making and the arrangement of symmetrical faces.

Plato took up the challenge to begin a more systematic study of solid three-dimensional shapes. In 387 BCE he set up an institution, the Academy, devoted to research and the teaching of science and philosophy. The visitors to the Academy were greeted by a sign that announced, "Let none ignorant of geometry enter here," a slogan that today adorns the entrance to the Mathematical Institute in Oxford. During discussions at the Academy, Plato's friend Theaetetus began to understand the principle that underpins the symmetry of shapes. Plato described his friend as having a snub nose and protruding eyes but a mind of beauty. And he would ultimately be remembered for his genius in capturing the abstract mathematics of symmetry.

Around the same time, early mathematicians recorded the discovery of another amazing symmetrical shape. Starting with five equilateral triangles arranged in a pyramid configuration, Theaetetus found that you could keep building such that one ensured that five triangles met at each new point of the shape as it evolved. It is a difficult shape to visualize, which is probably why it was the last symmetrical die to be discovered.

It must have been a wonderful moment for the first person who'd started to piece together triangles five at a time round a point, gradually building and seeing the shape evolve into a perfect shape with twenty triangular sides. It is not a shape that anyone had seen before in the natural world. Nor is its construction intuitive: one cannot piece together nineteen or eighteen triangles to create a shape where all the triangles meet in such a symmetrical manner. Only an intentional act of mathematical creation could bring this twenty-sided symmetrical shape into existence: the icosahedron.

This proved to be the last missing die. The ancient Greeks would go on to prove that there are only five possible dice that you can carve with all faces made up of the same two-dimensional symmetrical shape. This proof is the highlight of Euclid's great text, the *Elements*, which essentially launches the analytic art of doing mathematics. Once again, humanity's love for games and its wonder at mathematical discovery are intimately connected.

The mathematicians of antiquity had successfully created these dice, but it was not until the seventeenth century that the mathematics of how those dice behaved was finally worked out. In the intervening centuries, the roll of a die was considered unknowable and in the lap of the gods. It took two French mathematicians, Blaise Pascal and Pierre de Fermat, to wrestle the die out of the hands of gods and into the domain of man. They realized that even if an individual throw might be unpredictable, in the long run dice could be tamed. Probability theory, the mathematics of chance, proved revolutionary in both mathematics and the playing of games.

6: The Doubling Cube

THOUGH OUR TIME IN THE Middle East is almost at an end, we should double back one more time, as if we were playing one of the early racing games, to survey another ancient feature of humanity's love affair with games. If you open a modern backgammon board, then alongside the two dice used to race the counters around the board, you will find a rather curious third die. This die is responsible for another important component of playing games: gambling. Not simply content with winning a game, players were interested in raising the stakes of the games they were playing, risking money, possessions, and even their lives if they lost.

Gambling is found in ancient texts from the Bible to the *Mahabharata*, and to this day hundreds of billions of dollars are regularly staked on games of chance—yet this aspect of gaming has run into conflict with certain sections of society throughout human history. Gambling is inconsistent with many religious beliefs. Governments have prohibited their citizens from risking their livelihoods on the throw of a die. It is very addictive, and a self-destructive proclivity to gamble has been recognized by the *Diagnostic and Statistical Manual of Mental Disorders* as a psychological disorder in its own right. Simply put, gambling makes games dangerous. It could also be argued that if making money via gambling is the primary drive for playing a game, it no longer qualifies as a game at all but becomes instead a form of work.

Though ancient backgammon players certainly gambled, the doubling cube represents a surprisingly recent addition to the game. When backgammon reached the United States, American clubs added a new component to the game in the 1920s: the doubling cube, which features six powers of two on its sides: 2, 4, 8, 16, 32, and 64. At any point in the game, a player who feels confident of winning can take the doubling cube and challenge their opponent to double the wager. They do this by placing the side showing the two uppermost on the cube. If the player being challenged feels the game is unwinnable, they can refuse the double and forfeit the game for the agreed wager. However, once the challenge is accepted, the game continues with twice the prize at stake.

If the tide turns and the challenged player feels confident that they are now in a position to win, then they can take possession of the doubling cube and offer to double the wager again, turning the doubling cube onto the side showing the four. The game can ebb to and fro like this, with the gambling cube bouncing between the players, doubling in value each time. The doubling cube owes its success to the exciting, dynamic quality of backgammon that the lead can change dramatically multiple times during the course of a game.

To make successful use of the doubling cube, you need to be good at assessing the probability from a given position that you will win the game. Given that there is still a huge element of luck involved, since you don't know the roll of the dice, this requires a player with the skills to navigate a range of probabilities and uncertainty.

It is easiest to assess probabilities in the endgame when you have all your pieces in the home quadrant and you are trying to peel them off the board. For example, if you are left with two pieces still requiring a throw of six to get home, then four out of thirty-six possible throws of the dice will do this in one go: a score of 6-6, 5-5, 4-4, or 3-3. Remember a double allows you to move four pieces, so all these throws get you home. This means that there is a $\frac{4}{36} = \frac{1}{9}$ or 11 percent chance of winning in one throw. The probability goes up to 78 percent if you are looking to go out in two throws.

Some people insist that if your piece is three from the end, you need to throw a three to get it home. A four or higher isn't good enough for them. Don't play with these people. These rules make the endgame deadly tedious, upping the reliance on luck rather than strategy. All your good work getting your pieces into the home quadrant stands to be undone by unlucky throws of the dice. It robs backgammon of the qualities that make it exciting, bringing back the problems that make Senet such a tedious game.

If the endgame offers the clearest vantage point for assessing your chances of winning, it is in the mid-game, when so many twists and turns and interactions are still possible, that probabilities get harder to judge. This is an important skill if you want to challenge your opponent to double the stake. But what is the threshold at which you should offer a challenge or similarly turn down a doubling? The answer to this question is related to an important piece of mathematics that is used to understand random physical phenomenon: Brownian motion.

This phenomenon derives its name from its discoverer, the botanist Robert Brown. In the 1820s he first described this sort of movement when he looked under a microscope at the way pollen grains pinged about on the surface of water. They seemed to be moving in small steps randomly in different directions. Albert Einstein used this concept to give a convincing explanation that water consisted of discontinuous molecules that were acting like tiny billiard balls hitting the pollen in random directions. It was one of the major factors in convincing scientists of the existence of atoms as opposed to viewing matter as continuous.

Brownian motion can help to make predictions about the possible destination of a pollen particle as it pings around the surface. For example, in a square dish, Brownian motion helps calculate the chances that a pollen particle located at any particular point on the surface will end up at the top end of the dish as opposed to the bottom.

This can be useful for modeling backgammon. Think of the top of the dish as a win for black and the bottom of the dish a win for white. The position mid-game is like a pollen particle pinging about between the two scenarios. As the game proceeds, each throw of the dice acts like a little nudge toward either winning or losing the game. The point is to judge how close you need to be to winning the game for it to be worth betting on these random bufferings getting you to your goal.

In the simplest terms, applying Brownian motion to backgammon means knowing the point at which, if you are offered a double, accepting means that your expected payout should be greater than the expected payout if you refuse. To make good on my assurance that, if games are a way of playing mathematics, then mathematics should help you to play the game, what follows is a short exploration of the underlying math that can help a player assess when to accept a double in backgammon or to chuck in the towel and pay up.

To Double or Not to Double

Let's suppose you've bet y dollars on the game. Your opponent challenges you to double the stake. If you refuse the offer, you forfeit the game, and your expected payout is $-y$ dollars. Suppose there is a probability p that you can win the game from this point. The key is to determine the minimum value of p at which it is worth accepting the offer to double the bet and to continue to play.

There are two scenarios. In the first, you keep playing, increasing your chances of winning until they reach $1 - p$, at which point you are now in the position to double the bet again. At this point the other player has gone below the threshold to continue and caves. This has a probability of $\frac{p}{(1-p)}$ of happening and a payout of $2y$.

In the second scenario, you might carry on playing and lose without hitting this threshold. The probability of not hitting the upper threshold to double is then $1-\frac{p}{(1-p)}$, and the loss is $2y$. So the critical value of p is the one that solves the following equation:

$$-y = 2y\frac{p}{(1-p)} - 2y\left(1-\frac{p}{(1-p)}\right)$$

$$p - 1 = 2p - 2(1 - p) + 2p.$$

The solution to this equation is

$$p = \frac{1}{5}.$$

The algebra has spoken. Make your bets accordingly.

The bottom line is that if you still think you have a greater than 20 percent chance of winning when you are offered a double, it's best to accept. Similarly, if you think your chances of winning are over 80 percent, and you have the doubling cube, then it's worth offering to double at this point.

Our sojourn in the Middle East is at an end, and our next destination on our journey around the board is a country that deserves special honors as the home of a great many games: India. But in math, games, and life, the pleasure is in working things out along the way—so let us take a moment during our voyage across the Arabian Sea to contemplate what drives us to play games in the first place.

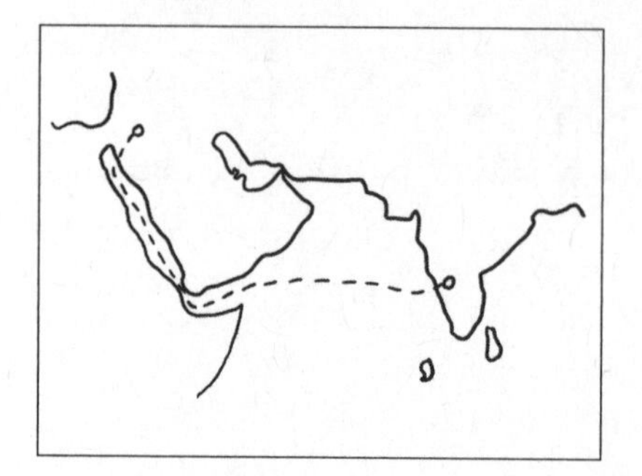

CHAPTER 2

THE ARABIAN SEA

What Is a Game and Why Do We Play?

WHEN PHILEAS FOGG SET sail aboard the *Mongolia* to make his way to India, Jules Verne described for the reader the way that such an "eccentric personage" passes his time on a voyage: with good food and the playing of many games. "He made his four hearty meals every day, regardless of the most persistent rolling and pitching on the part of the steamer; and he played whist indefatigably, for he had found partners as enthusiastic in the game as himself."

Our journey through the world of games won't feature quite as much downtime as Fogg's, but the time passing from one continent to another offers us an opportunity to think afresh about the games we play. As we make our way from the Middle East across the Arabian Sea to India, I want to address the question what is a game and why do we play?

7: *Homo Ludens*

MANY THEORIES HAVE BEEN proposed about the role of play in the evolution of the human species and about how integral the act of play is to the human condition. As eighteenth-century playwright and philosopher Friedrich Schiller wrote, "Man only plays when in the full meaning of the word he is a man, and he is only completely a man when he plays." Crystallizing this, Dutch cultural historian Johan Huizinga suggested, famously, that humanity was so devoted to play that we should refer to ourselves as *homo ludens* rather than *homo sapiens*. A key player in the anthropology of games, Huizinga argues in his 1938 book *Homo Ludens*, still regarded as a seminal text, that our species owes its evolution to the power of games.

His work wasn't just dedicated to games. Following the publication of this book, he became very critical of fascism and the German occupation of the Netherlands, resulting in his detention in 1942 by the Nazis. He died just a few months before the liberation of his country. But his work on the role of games in human evolution has ensured him a place in history.

Importantly, Huizinga admits that there are other species that engage in play. Plenty of animals engage in play fighting, abiding by the rule that they should hold back on really biting and hurting their opponents. An evolutionary biologist might argue that play represents a training ground for young animals before they encounter similar scenarios in their adult lives. And it is perhaps striking that in the animal kingdom play is generally the preserve of the young. In contrast, many of the games I have encountered on my journey round the world are aimed at adults as much as or even more than at kids.

Seeking to answer the question of why we play games, Huizinga cautions against falling into the trap of seeking an explanation that serves something outside simply playing the game. Yes, perhaps the act of play teaches us something. But is that really why we set out to play games? Not according to Huizinga: "Play is a free activity standing quite consciously outside 'ordinary' life as being 'not serious,' but at the same time absorbing the player intensely and utterly. It is an activity connected with no material interest, and no profit can be gained by it. It proceeds within its own proper boundaries of time and space according to fixed rules and in an orderly manner."

In other words, we play for fun—a motive that seems to defy any attempt to analyze, intellectualize, or interpret it in a logical, rational manner. Fun cannot be reduced to more basic concepts when it is already atomic. And if fun is not present—if, for instance, we're playing chess to train ourselves for warfare or playing poker to earn money—then we have ceased to play a game at all and put ourselves to work.

One aspect of play that connects it uniquely to our species is its relationship to the brain. To play implies a theory of mind: the ability to summon up imaginary scenarios, to work within self-imposed rules that are followed in the minds of the players, to conjure up a world that works in a different time frame. These all require a brain that can imagine other worlds, project forward into the future, and recognize the minds of others against whom it is pitting its wits.

Irving Finkel, the historian at the British Museum who decoded the rules of the Royal Game of Ur, has speculated that the invention of games pitting one opponent against another happened in tandem with the development of human consciousness. Playing a game offered a way of recognizing that the other members of your group did not share your conscious experience. It is my belief that human creativity emerged as part of our tool kit for exploring consciousness, a view also held by psychologist Karl Rogers, and I think Finkel is right to identify game playing as a tool that our distant ancestors developed to examine the mind of the other.

Though game playing has a lot to do with competition against the other, there is an intimacy to the act of play as well. Even after engaging in a pitched contest against an opponent, players often emerge from the game back into real life with the sense that they have been bonded closer together by the experience. The game therefore acts to fuse a group of distinct consciousnesses through common experience. This helps explain why we often prefer playing games with others rather than alone. Even better if we can show off our prowess at a game before an audience. Despite the proliferation of online gambling, many people still prefer throwing the dice on the craps table in Vegas in front of other punters to gambling in solitude on their home computers.

Tellingly, internet gambling can begin to feel more like work than play, since the imperative to make money is all that is left once you remove the public element of enjoying a game in the company of others.

One of the reasons Huizinga believes we are drawn to games is the perfection they offer as an alternative to the chaos of real life. "Into an imperfect world and into the confusion of life it brings a temporary, a limited perfection," he writes. The confined space and time of a game allow us to escape the mess and injustice of the world around us and create something beautiful, even if it is temporary. This reasoning could just as easily describe my attraction to mathematics. Both games and math provide me with perfect, beautiful rule-bound worlds that allow me to escape the reality around me. As French poet Paul Valéry expressed, "No skepticism is possible where the rules of a game are concerned, for the principle underlying them is an unshakeable truth." In defiance of Ludwig Wittgenstein's claim that it is impossible to define a game, Huizinga offers an attempt to pin the concept down: a game is time bound; it has no contact with a reality outside itself; its performance is its own end. It is sustained by the consciousness of being a pleasurable, even mirthful, relaxation from the strains of ordinary life.

There are certainly parts of our culture that might pass the test to be admitted as a game we play, and Huizinga is interested in exploring what this tells us about ourselves. Poetry fails to pass the test of a game because the words lift the poem into ideation and judgment. Music, on the other hand, never leaves the play sphere. Observe how we even say we play music using the same verb that we use for engaging in a game.

An argument could be made that science might therefore be admitted as a game. Given that scientists are isolated within their own field and bound by strict rules determined by their methodology, Huizinga wrestles with the "amazing and horrifying conclusion" that science might deserve to be included in play. But the fundamental connection between science and reality allows him to step back from the brink. This leaves me to wonder about whether the mathematics that I engage in also qualifies. It can't be saved from play by appeal to connections with reality. I literally create independent worlds that have no physical basis, with internal rules, that are outside time, and that I principally create for the joy of it. I think Huizinga might find it harder to exclude mathematics from the realm of play. Whether he would still be horrified by this classification is unknowable—I would agree with him that it is amazing.

Huizinga's work was foundational, but he also had his critics. French intellectual Roger Caillois, in *Man, Play and Games*, published in French in 1958, attempted his own list of the defining qualities that make a game and criticized Huizinga's view that play is atomic and resistant to breaking up into components. He was far less credulous about the unknowable nature of fun—for example, when a colleague showed him a Mexican jumping bean, Caillois insisted that it wasn't magic and that if you cut it open, you'd likely find a worm or larva inside causing the bean to jump in the air. Though he may have been right, his colleague was horrified at Caillois's insistence on destroying the poetry of this toy by demanding an explanation and unweaving the rainbow.

Caillois was never content with leaving things unexamined, and his analysis of games tries to pull apart what constitutes the definition of this strange activity. He ultimately identified six key traits: freedom, separation, uncertainty, unproductiveness, rules, and the imagination.

First: A game demands freedom. Anyone who is forced to play a game is working rather than playing. A game therefore connects with another important aspect of human consciousness: our free will.

Second: A game is separate. This refers to the quality that Huizinga identified where a game should operate outside normal time and space. The game should have its own demarcated space in which it is played with a set time limit. It should have its own beginning and its own end.

Third: A game demands uncertainty. We enter the game because there is a chance either side will win, and if we know in advance how the game will end, it loses all its power. That is why ensuring ongoing uncertainty for as long as possible is a key component in game design.

Monopoly suffers as a game because once someone has gained the upper hand, it's hard to turn the tables. In games of skill, it is also important that opponents are matched in ability. Gary Kasparov against Donald Trump would not make for a chess match that holds much uncertainty, though a game of snakes and ladders might be a more even contest. That is partly why we are so drawn to games of pure chance, because they allow those without any skills the opportunity to win.

Four: A game is inherently unproductive. This picks up on another quality that Huizinga identified as being a key component of playing games. They have no point beyond perhaps satisfying the human desire to win. They are not created as teaching tools. Their aim is not to create wealth even if the side effect of betting on games can accomplish that outcome.

Five: Games are governed by rules. Wittgenstein too felt that this quality was absolutely crucial to games. It is key to his concept of language games that by using words we reveal the rules they obey.

Six: Games are make-believe. A game consists of creating a second reality that runs in parallel with real life. It is a fictional universe that the players voluntarily summon up independent of the stern reality of the physical universe we are part of.

Pulling all of these threads together, why do we really play games? According to Caillois, it is "an occasion of pure waste: waste of time, energy, ingenuity, skill, and often of money." This may seem a damning commentary on how we spend our time when we play games, but if we start to analyze why we do just about anything in the face of mortality, we're likely to reach the same conclusion. There is no innate point to our existence. Life has no meaning. Our species will get wiped out in the not-too-distant future. And we are doing a good job of accelerating that end. Our high level of consciousness, however, is something probably unique to our species, and it allows us to do something many other animals cannot: mental time travel. Even so, the cost is high. It means we can contemplate our death: both as individuals and as a species.

We have developed coping strategies. One is to falsely manufacture meaning through things like religion. But perhaps games are another clever coping strategy. By simulating the waste of time that a meaningless life seemingly entails, games become a microcosm of what we must endure as we play out the game of life on the macro scale. We read a novel not to teach ourselves how to live but principally for fun. Much as it may allow us to examine our inner worlds, to empathize with the inner worlds of others and to realize we are not alone, isn't reading really about the joy it brings? The same goes for the games we play: we play them because they are fun, bringing joy to a life that would otherwise give us little reason to make a next throw of the dice.

It is interesting to speculate that although games probably evolved with our species' graduation into conscious, independent beings some two hundred thousand years ago, the shift to the elaborate rule-based worlds that are defined in different arenas of time and space might coincide with the recognition that our own universe follows rules. The ancient Greeks and the Babylonians were just beginning to understand patterns in the movement of the stars and planets. These were mirrored in their games, where the pieces follow trajectories around the board bound by the rules of the game. The Royal Game of Ur could easily be seen as a model of planets passing through the constellations of the night sky, an early form of orrery.

Games also emerged at a time when human society was introducing rules to guide its progress. A democratic society is one that agrees to an equality of each player under the political rules of the game rather than submitting to a tyrant's whim where the dice are loaded and the game unfair.

But for all that games teach us about human history and consciousness, we must also bear in mind that our species is not alone in its fondness for play.

8: Animal Games

ELEPHANTS SLIDE DOWN MUDDY embankments knocking each other over along the way. Groups of thirty meerkats play at fighting in a frantic ball of fur and legs. Coyote pups pounce on each other in mock attacks. Bees get a buzz out of playing with balls. Dolphins blow and chase rings of air under the sea.

As Charles Darwin wrote in *The Descent of Man*, "Happiness is never better exhibited than by young animals, such as puppies, kittens, lambs, and company, when playing together, like our own children." But why do they play, and does it have anything to do with humanity's fondness for games?

The traditional theory holds that young animals play in order to hone skills for adult life. Yet it is not clear how much impact childhood play really has on the future path of animals. Research into whether more playful youngsters make better hunters has revealed that there doesn't seem to be much advantage visible later in life.

Domesticated animals spend a lot of time playing well into adulthood. Dogs will chase and retrieve a ball endlessly. Cats enjoy pouncing on a stuffed mouse on the end of a stick even though they know it isn't food. They understand the rules of the games. No longer needing to work to survive, are these animals just passing the time in their private utopias? Or else are they simply simulating the playful attitude of their owners? But even outside the domestic realm, wild animals still seem to enjoy playing with balls and chasing sticks—that would seem to rule out the influence of leisure and the example of humanity.

Frankly, animal researchers remain rather mystified by the fact that animals play. After all, it is expensive in evolutionary terms. It puts them at unnecessary danger of injury. It makes them vulnerable to attack. Surely there must be some advantage to play, but it is hard to identify.

Some of the most interesting progress on an explanation for time spent playing has been unearthed through research on rodents. Rats denied play in early development were less prepared to deal with stressful scenarios in later life. They would lash out or hide in a corner, while the rats who had been reared on play were able to moderate their response to stress. Play therefore seems to wire the brain to release stress outside the realm of play.

There is also evidence that playing can lead to a larger brain. Rats reared in a rich environment of games had larger cerebral cortexes with increased neural connectivity. Recent research on a sample of Australian native birds has confirmed the role of play in brain development.

The use of tools was not sufficient to increase brain size alone—it genuinely seemed to be the element of play that had an effect.

But is there any evidence of animals creating the kinds of rule-bound games that humans have fashioned? Play fighting seems to come along with rules to limit the amount of damage you can inflict on your opponent, holding back from really biting off your sister's ear. There is interesting evidence of willow wrens marking out an arena for fights with rival wrens. An area of elevated ground covered in short grass and about two meters in diameter will be nominated for the contest, and adult birds will gather in the arena to await rivals. The rule of combat is that you are not allowed to leave the fighting ground, as if the wrens were playing under the rules of sumo wrestling, trying to push their opponents out of the ring. This is the board on which this game is played.

But there is little evidence of games that go beyond mimicry of adult scenarios. The creativity and imagination necessary to transport an animal outside its natural surroundings into an imaginary world do not seem to be in evidence. Perhaps what makes the difference is another thing that separates humanity from other species: language.

9: Language Games

WE HAVE ALREADY TOUCHED upon language games briefly in discussing a philosophical proposal described by Ludwig Wittgenstein in his 1953 treatise *Philosophical Investigations.* His key suggestion is that words only gain meaning by how they are used. We understand their meaning by witnessing how the rules of the language game are implemented.

One of his principal examples is how we understand the very word *game.* What is a game? Is it a competitive activity with rules played for entertainment? Wittgenstein argues that any attempt to write down conditions to define a game will inevitably miss some things that we regard as games and embrace others that fit the definition but that we didn't mean to include as a game.

"Consider for example... board-games, card-games, ball-games, Olympic games," Wittgenstein writes. "What is common to them all?—Don't say: 'There must be something common, or they would not be called "games"'—but look and see whether there is anything common to all.—For if you look at them you will not see something that is common to all, but similarities, relationships, and a whole series of them at that. To repeat: don't think, but look!"

We understand what a game is without needing a definition—but how? Wittgenstein argues that we understand what the word *game* means by the way we use it, through the examples we give, by explaining why this counts as an example of *game*, and by providing unforeseen examples of games that defy attempts to define the word. All of these activities are part of playing the language game. Words, he believes, gain meaning by a process of family resemblance.

Wittgenstein uses a nice image of how we build up the concept of a word like *game*: "As in spinning a thread we twist fibre on fibre. And the strength of the thread does not reside in the fact that some one fibre runs through its whole length, but in the overlapping of many fibres."

One might argue that our not being able to write down a satisfactory definition of *game* doesn't mean that such a definition doesn't exist. For Wittgenstein, however, there is a deeper importance to concepts like *game*, which are understood while nevertheless resisting any explicit definition. The quality of vagueness and fuzzy boundaries on display here is important in creating language. And, Wittgenstein concludes, only by playing language games can we understand these fuzzy edges.

Wittgenstein came up with his idea of language games while watching a game of football. He realized that to anyone who didn't understand the rules, a game of football would look like a chaotic, meaningless activity (although my wife does understand the rules and still thinks that the game is chaotic and meaningless).

It is for this reason that artificial intelligence (AI) finds it very difficult to deal with language. Any top-down attempt to encode language by defining meanings of words is doomed to failure. The bottom-up approach of machine learning has been much more successful because it works by playing the language game to a certain extent. The extraordinary power of natural-language generators like GPT shows how successful this approach has been. By encountering more and more examples of the use of the word *game* in context, it can hope to build up an understanding. But the language game needs to be embodied as much as just limited to language itself. And so, without embodying the experience of enjoying a game or feeling the tension of uncertainty in the outcome, the AI is missing an important part of the language game.

Among the language challenges that so often catch out AI are *Winograd challenges*. Take the following sort of sentence: Alan can still beat Bob at tennis even though he is thirty years older. Who is older? Alan or Bob? Why do you know immediately that the answer is Alan? The AI finds the ambiguity very difficult to parse. You need experience of tennis to realize that age might slow you down. If tennis was changed to chess, it might be an advantage to be older, especially if Bob was a child.

AI is getting better at these sorts of challenges. A 2020 paper from OpenAI, the creators of GPT-3, reported that it was getting 90 percent of these ambiguous Winograd challenges correct. But it seems to be achieving this sort of success rate via statistical shortcuts: do a google search and analyze all scores of games of tennis on the internet and see if older people are more likely to win. This still falls short of true understanding.

Wittgenstein's insistence on games' incomprehensibility not only makes them impervious to AI but puts our entire journey through this book to the test. Is there really no way to understand a game other than by playing it? You might understand the rules of a game like Monopoly, but is it only in the embodied action of playing it that you can truly know the game?

10: The Grasshopper's Games

NOT EVERYONE IS DEFEATED BY Wittgenstein's claim that games can't be defined. Philosopher Bernard Suits wrote a treatise about his definition of a game, fending off all attempts by Wittgensteinians to poke holes. His definition is wonderfully succinct: "Playing a game is a voluntary attempt to overcome unnecessary obstacles."

Based in the philosophy department at the University of Waterloo for much of his working life, Suits was celebrated by students as a fantastic lecturer. As one student wrote, "Professor Suits' classes usually spend the first few days of term being moved to larger and larger rooms." His book, *The Grasshopper: Games, Life and Utopia*, not only discusses the philosophy of games but captures something of the performances he must have given in his lectures. His defense of his definition takes on a wonderfully playful character, truly in the spirit of the games he is determined to define.

The 1978 text takes the classic Aesop's fable "The Grasshopper and the Ant" as its inspiration. In the fable, an ant berates a grasshopper for spending all its time playing during the summer when it should have been working hard to put aside provisions for the tough winter ahead. The ant survives because of all the hard work it has done preparing, whereas the grasshopper dies. The moral is that hard work now is rewarded in the future, and Aesop clearly sees the ant is the insect we should aspire to mimic.

Suits takes the opposite view. Following Plato's lead, his book adopts the form of a dialogue between the grasshopper and a range of other protagonists who try to challenge the grasshopper and Suits's take on games.

Pushing the book's Platonic inspiration even further, the grasshopper is shown as taking the role of Socrates in the dialogue. Just as Socrates turned down the chance to live at the expense of continuing philosophical debate, the grasshopper refuses to accept food that will help him survive the winter if it means betraying his belief in the power of playing games.

Suits believes that playing games will be the ultimate goal of a utopian society that has removed the need to work. Such a view is consistent with Karl Marx's description of the highest form of communism as a society where work has become a free choice. In such a society, individuals choose to place unnecessary obstacles in the way of carrying out an action only because they enjoy the challenge of overcoming those obstacles.

Even in our not-yet-utopian society, this joy in doing things the hard way can be found everywhere in human behavior. Edmund Hillary would still choose to climb Everest the long way even if there were an elevator he could take to the top on the other side of the mountain. There is a satisfaction in cooking your own pavlova rather than simply buying one at the supermarket. And many people opt to work through mathematical problems and puzzles even when they could look up the answers in the back of the book.

Work will become a game in Suits's view of utopia. Some will choose to build houses with their hands rather than getting the machines to do it for them. Suits even argues that the creation of art will not survive true utopia since art is the struggle to understand our place in the world, and the removal of that struggle will foreclose on art's reason to exist. In utopia human hopes, fears, triumphs, and tragedies will have disappeared. In the absence of work and art, all that will be left is the act of playing games, the last remaining supreme human good.

Suits's definition of what constitutes a game involves three parts. The game should have a goal that players are trying to achieve—perhaps it is to maneuver your opponent's king into checkmate. The game should also have rules that restrict the ways that you can achieve that goal, so that you can't just lift up a selection of your pieces and surround the opposing king at will. The final piece of the jigsaw is the voluntary acceptance of these rules. And the acceptance should be granted for the pleasure of playing and passing the time, not for any external value. It is the process, not the product, that brings the pleasure.

With the grasshopper's definition established, Suits goes on to document attempts by various protagonists to undermine the ideas on the table. Some try to come up with things that they believe fit the definition but shouldn't be called games; others list games that don't meet the criteria. Each time the grasshopper successfully defends his pitch. As philosopher Thomas Hurka describes Suits's project, it is "a precisely placed boot in Wittgenstein's balls."

Having explored why we play and defined as best we can what we mean by a game, we are ready to throw the dice and move our counter to grapple with the games of our next destination: India.

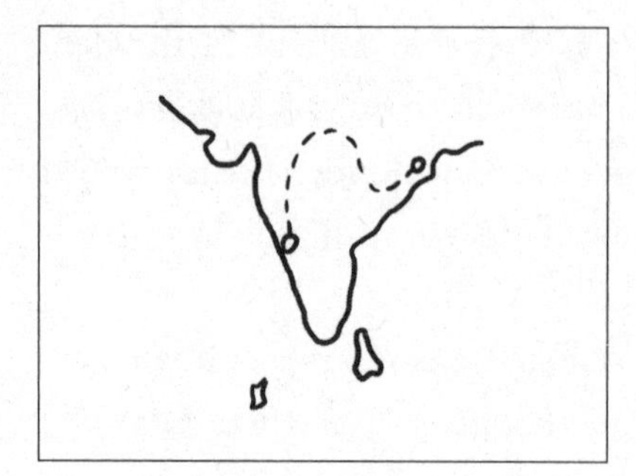

CHAPTER 3

INDIA

OVER THE COURSE OF MANY VISITS, I have found India to be a country that loves its games and reveres mathematics. Here being bad at math is no badge of honor, like it is in the West. Many great mathematical ideas traditionally thought to be European in origin actually started in India: calculus, the concept of infinity—even the Fibonacci numbers were discovered in India centuries before the Italian mathematician got his name on them.

But India is also home to some of the most iconic games of all time. If the Middle East gave us one of the best racing games, then India is the birthplace of one of the greatest war games that humans have invented: the game of chess.

11: Chess

THE MARK OF A REALLY GOOD GAME is when a match is still being talked about 150 years after it was played. The Immortal Game is one such standout, still exciting contemporary players for its drama and audaciousness. It was played in 1851 between the champion of the day, Adolf Anderson, and Lionel Kieseritzky, a chess teacher who whiled away his days playing chess in the Café de la Régence in Paris. The setting was the very first international chess tournament held in London, which ran in parallel with the Great Exhibition of 1851 showcasing British technology and culture.

Kieseritzky had already lost to Anderson as part of the official tournament, but the two agreed to play a few friendly games to entertain themselves before Anderson moved on to his next rounds. It was one of these informal games that has now gone down in gaming history as the most extraordinary match of all time.

In most games of chess, a player seeks to hold on to the powerful pieces on the board, like the queen and the rooks, because they are the most valuable in achieving checkmate. So Kieseritzky was staggered by Anderson's decision to sacrifice a bishop after eleven moves, then both his rooks and his queen in the following ten moves. Even at this seeming disadvantage, Anderson went on to pin Kieseritzky's king in checkmate on his twenty-second move using only the minor pieces he had left on the board. Kieseritzky himself had only lost three pawns up to this point.

Kieseritzky was shocked. That his opponent's decimated army could yet achieve a win with such depleted forces was beyond belief. He immediately wrote down the moves and sent them to his friends in Paris, where the game was published and named the Immortal Game by chess journalist Ernst Falkbeer. It represents a classic example of the powerful narrative quality inherent in a good game. The abstract chess notation "Nxg7 + Kd8 22. Qf6 + !" hides inside it a story with drama as boundless as that of a great Verdi opera. It is like musical notes on the page that, once given voice, transform from dots of ink into the keening sounds of pure emotion.

One reason that chess has been such a successful game is the delicate combination of constraint and freedom that allows for such a wide range of outcomes. Like a good story with its classic three-act structure of beginning, middle, and end, a chess game too has an opening, a middle where the game develops, and then the dramatic endgame concluding with a checkmate.

The Immortal Game is an example of a whole new style of games played during the mid-1800s, known as the Romantic period, and at times it can seem like the game is taking cues from the music of the era. Anderson and Kieseritzky are the Brahms and Tchaikovsky of the chess world. As Marcel Duchamp observed, "I have come to the personal conclusion that while all artists are not chess players, all chess players are artists."

There is an apocryphal story that mathematicians love telling about the origins of chess because it illustrates one of the most important concepts of mathematics: the power of exponential growth.

Legend has it that the game was invented for a Persian king by one of his advisors. The king was so enamored of the game that he promised the advisor great riches as a reward for his creation. The king was rather taken aback when the advisor simply asked for one grain of rice to be placed on the first square of the chess board, two on the second, four on the third, and to continue like this, doubling the number of grains from one square to the next. The king immediately agreed thinking he'd gotten off lightly. After all, along the first row of the chess board there was a total of 255 grains of rice—barely enough for a small piece of sushi.

What the king hadn't taken into account was the power of doubling. Already by the halfway point on the board, the king needed 280,000 kilograms of rice to meet the advisor's request. And this was still the easy half of the board. To cover the whole board, doubling the grains at each square, would require a total of 18,446,744,073,709,551,615 grains of rice. That's more rice than has been produced on our planet in the last millennium. The king failed to meet his advisor's request and, as legend has it, had to forfeit his kingdom instead.

It is a charming story that illustrates the potentials of exponential growth, though there is no evidence that it was based on a real-life event. But there is perhaps a grain of truth in another part of the story. We know that a version of the game was presented as part of a tribute sent from the king of India to the Persian court around 600 AD. An epic poem of 120,000 lines describes how the Indian king's ambassador challenged the advisors to the Persian ruler to "solve the mysteries of this game. If they succeed my master the king of Hind will pay you tribute as an overlord, but if you fail it will be proof that the Persians are of lower intellect and we shall demand tribute from Iran."

It seems that they were required by the next morning to come up with the rules of the game rather than to actually play a game against the Indian ambassador. This was a rather meta challenge, which apparently one of the advisors was able to meet, although, as game expert R. C. Bell suggests, "Perhaps the 24 hours were spent in bribing the Indian ambassador rather than in heavy thinking."

But chess didn't arrive fully formed. It is a good example of the way some games must change and mutate before evolving into the superior beasts we see at play today. Any modern-day game developer will never release a new game to the market without a testing process where teams of players put it through its paces. When it comes to chess, Dutch biologist Alex Kraaijeveld even applied a phylogenetic analysis, usually reserved for tracing the origins of different animal species, to determine the game's likely origins. A comparison of forty different early versions of the game placed on an evolutionary tree gives a good indication of the origins of this wonderful beast, and it seems that we should indeed head to India for the first signs of chess emerging in the kingdom of games.

The true origins of chess are believed to hearken back to a game called chaturanga in India. Surprisingly, it seems this was actually a four-player game rather than a two-player game. The word *chaturanga* is Sanskrit for "four divisions," and the game involved four armies, colored red, yellow, green, and black, positioned on the four edges of the square board. The board itself was originally used for another game called Ashtapada, a racing game a bit like Ludo, but now repurposed for this new game of war.

Each army consisted of an elephant, the early forerunner of the rook; a horse, which already moved like the modern knight; a ship or chariot, which would evolve in Europe into the bishops; and the pawns. The rajah, like the king in the modern game, was the target of the other armies. It seems that during the game, a player could capture another player's army and would therefore have two of each of the dominant pieces, which explains why the modern game sees an army mirrored on either side of the king and queen.

The game began to be known as shah once it moved to Persia, as this was the Persian for "king," and it seems that the word *chess* evolved from the Persian name. At this point in its evolution, the roles of the chariot and the elephant got transposed so that the chariot began behaving like the rook, moving orthogonally across the board, while the elephant moved diagonally. As the game became more popular in the Arab world, we also see the pieces changing from ornate physical versions of elephants and chariots to more abstract representations. This was partly a result of the Quran's prohibition on creating images of living beings but also mirrors the Arab world's development of abstract algebra, where numbers are replaced by abstract letters. It is in this abstraction that we can explain how the elephant would eventually transform into a bishop in the European game. The abstract piece for the elephant had a cut in the top of the piece, meant to give an impression of tusks, which the European players believed looked rather like the miter worn by the bishop. In time, this resemblance produced the chess piece that we recognize today.

A revolutionary innovation established by chess is the different roles that pieces play in the game. Never before had a game considered pieces moving in different ways. This innovation was key to the rich nature of game play. But how these pieces moved would take time to evolve. Originally the bishop could only move diagonally two squares at a time. The restricted nature of the pieces meant that often players would be allowed ten opening moves each to rearrange their army on their side of the board before turns were taken as the armies engaged.

Up to this point there was no queen. An advisor was eventually introduced with limited movement, but the evolution of the piece from advisor to queen, equipped with the most sweeping moves of any piece on the board, seems to have happened only once the game reached Europe via the Arab world and perhaps reflected the growing political importance of women.

It's interesting to speculate whether, as the nature of war changed over the centuries, this also had an effect on the game of chess. In the past battles were fought in close proximity, and soldiers were limited in how fast they could move. But the invention of new weapons, like the long bow and musket, meant that soldiers could engage over much greater distances. The shift from the bishop moving in small steps to a piece that can sweep diagonally across the board came at a time when these new weapons were being used in warfare. Is the bishop's reach a reflection of the shift to long-range warfare? Some people even refer to a bishop sitting on the edge of the board as a sniper.

The other rather curious discovery that I came across is the original role that dice may have played in chess's ancestor games. A rod-like die that could land on one of four sides would be used to determine which piece the player would use on his next move. But when gambling and games of chance were outlawed in Hindu culture, players of the game found a way to evade prosecution by simply taking away the element of chance and allowing each player to decide which piece to move. The game went from one of chance to the game we recognize today, which hinges on the intellect of the player and nothing more.

Writing in 947 AD, Arab mathematician Al Masudi recommended chess over backgammon precisely because chess is an expression of humans' free will, while backgammon instills in the player a sense of fatalism. In chess the better mind wins. Al Masudi was also the first to document the story of the exponentially increasing rice. The mathematician in him couldn't resist the chance to explain to his readers the power of exponential growth.

A sign of how popular the game quickly became in European circles, once it had arrived via the spice route, is the fact that the second book to be printed on William Caxton's press was one about chess. Published in 1475 *The Game and Playe of the Chesse* explains how the pieces move on the board but also uses the game as a metaphor to sermonize on the state of fifteenth-century society.

The game even makes its way into Shakespeare's *The Tempest*. At the end, Ferdinand and Miranda engage in a game of chess, and their dialogue while playing is fascinating. It consists of eight lines of eight words each—a chess board embedded in the dialogue of the play.

Chess is often regarded as the mathematician's game of choice. As famous Cambridge mathematician G. H. Hardy remarked, "Chess problems are pure mathematics, they are the hymn tunes of mathematics." Seeing the logical consequences of each possible move and tracing the tree of possibilities into the future are trademark mathematical ways of thinking. But there is an aspect of the game that I find rather unmathematical, and it certainly has been key to why I've never been terribly successful at playing the game: openings.

Openings are so crucial to chess that a whole literature is devoted to the early part of the game. If white moves the queen's pawn as the opening move, what is the best response? These opening sequences are well known by seasoned players of the game: the Queen's Gambit, the Sicilian Defense, the Spanish Opening. The *Oxford Companion to Chess* lists 1,327 different named opening sequences.

The trouble is that if you are playing an opponent who knows their openings and you don't, by about ten moves into the game, you've most probably already put your army at a strategic disadvantage. The requirement to master these openings has always challenged my rather poor memory. I think that I was always particularly drawn to mathematics because you didn't actually need to remember stuff—I could always just reconstruct whatever I needed from first principles.

Though Kieseritzky lost the Immortal Game to the extraordinary sacrifices that Anderson made after opening with the King's Gambit, he was also famous for his openings. Many still bear his name, such as the Kieseritzky Gambit, the Kieseritzky Attack, and the Boden-Kieseritzky Gambit. He even experimented with variants of the game, inventing a three-dimensional version played across eight boards stacked on top of each other to make a huge 8 × 8 × 8 playing arena. The game didn't really catch on, though a modified version was featured during several episodes of *Star Trek.*

Chess is not the only game to mimic warfare: Battleship, Risk, L'Attaque, the Roman game of Ludus Latrunculorum, and numerous other games all use war in their marketing. But why is it that warfare, a deadly serious, dangerous activity, and game playing, a fun, relaxing, enjoyable pursuit, should share so much in common? Some have theorized that war games offer a safe space to try out strategy. They can sharpen the mind of the leader of a campaign to understand the consequences of certain moves in the field. In the early nineteenth century, the Prussian high command used a game called Kriegsspiel, German for "war game," to test the abilities of aspiring officers to analyze military strategy. The Prussian successes against France that ensued persuaded other countries to adopt the game as a training ground for the military.

Alternately, the converse explanation holds that games of war are a way of avoiding real war by allowing the game to be the space where aggression and conflict can be released. American psychologist William James has suggested that war has a positive psychological effect that we crave. He argues that it provides a society with a sense of cohesion and moral purpose that transcends the monotony of everyday life. But couldn't a game provide the same effect? The human species seems to have a need to let off steam, and it's all too common to respond by picking fights that we can't win. War games may offer a less costly way of meeting these underlying psychological needs.

Games have even been used explicitly to stand in for violence. While traveling in central America, I learned about a ball game used by the Mayans as a substitute for outright conflict between two communities. Medieval warfare was sometimes decided by two knights fighting to the death to avoid whole armies being slaughtered. The chess matches between the American Bobby Fisher and the Soviet Boris Spassky allowed these superpowers to play out the tensions of the Cold War in a game. Games provide a way for humans to transcend the pathology of war.

Although games are entered into freely and often with the goal of having fun, once immersed in the game, players compete with a seriousness and commitment that rivals military engagement. A game is set up precisely to pitch players against each other in competition with no punches pulled. The crucial difference is that games are an end in themselves, whereas war is a means to an end. At the end of the game, enemies become friends, bonded by the experience of sharing the game together.

Still, conflict is central to the best games. The idea of a collaborative game where players work together to win collectively is a remarkably modern concept that would probably be quite alien to those who played games in ancient times.

Warfare, like games, has rules that the two sides abide by. So perhaps they find themselves connected not because games help to train soldiers but simply because the framework of games shares much in common with the schema of war.

12: Carrom

MANY OF MY TRIPS TO INDIA have been mathematical pilgrimages to seek out the origins of some of the most important ideas that began there. Though it seems incredible to think that zero needed to be invented, prior to its discovery, numbers existed to count things, and there was no apparent use for a number to record an absence of things to count, or nothingness.

During the seventh century, in a stroke of brilliance, Indian mathematicians, in particular a mathematician called Brahmagupta, recognized the power of counting nothing. They may have been assisted by Hindu philosophy, in which the void was never just nothing but was instead the possibility of something. Perhaps this is why Indian thinkers were not frightened to embrace the idea of zero, while Europeans found the concept frightening. When the Arabs finally brought the idea to Europe in the twelfth century, zero was promptly banned as the number of the devil.

I once traveled to India to visit a small temple that hangs off the side of a cliff in the fort town of Gwalior. Dedicated to Vishnu and built in 876 AD, the temple can barely fit two people inside. Inscribed on the walls of the temple I found one of the first recorded uses of this new number zero. Its discovery would ultimately unleash exciting new mathematical insights that had been hidden without the ability to count nothing.

But it wasn't just the temple that I fell in love with. The town of Gwalior is a mathematician's paradise. Usually billboards entice passersby to buy Coca-Cola or the latest mobile phone. Not in Gwalior. I have never seen so many posters advertising math lessons. Everywhere you looked there was another billboard offering to take you on the path to numerical nirvana. My favorite was an advertisement for a Mr. Pramod Kushwah declaring that math is "one more reason to believe."

As always, I was keen to track down any new games that I might not have encountered before. What were the locals playing on the streets of the city? It was in Gwalior that I encountered my first carrom board.

A kind of combination of billiards and air hockey, the game is played on a square board about seventy-five centimeters square with four pockets, one in each corner, into which you are aiming to fire the carrom pieces. These consist of flat wooden discs rather like those used to play draughts. Known as carrom men, these include nine black and nine white pieces, plus a single red piece called the queen. At the beginning of the game, they are arranged in the middle of the board a bit like the fifteen balls in snooker or pool.

What immediately caught my interest was the arrangement of these nineteen pieces. At first nineteen seems like a strange number of pieces to arrange in any coherent pattern. But it is actually an example of something called a *hex number*. Place the queen at the center of the board, then you can surround the queen with six pieces. Around this it is possible to put another layer to enlarge the hexagon using the twelve remaining pieces of the game. The pieces can be arranged in alternating colors. We will see this configuration coming up again when we build the board behind one of great modern board games, the Settlers of Catan.

This arrangement is the beginning of the hexagonal packing of two-dimensional space by circles. It is the most efficient way to pack circles, wasting only 9.31 percent of space. This was only proved in 1942 by Hungarian mathematician László Fejes Tóth. A symmetrical arrangement in what mathematicians called a *lattice* was proved to be the best by French mathematician Joseph Louis Lagrange in 1773, but Tóth showed that no weird asymmetrical arrangement could beat the beehive structure.

The Math of Carrom

The hexagonal orientation of the carrom board is part of what makes the game so mathematically fascinating. If we continued the hexagon, expanding by a new layer each time, then the number of carrom men needed follows the following sequence:

1, 7, 19, 37, 61, 91, 127...

The following formula calculates the number of men needed as the hexagon expands:

$$3 \times n \times (n + 1) + 1.$$

If you add up the hexagonal numbers, then you get the following sequence: 1, 8, 27, 64, 125, 216... Do you notice anything unusual

about these numbers? Rather curiously you always get a cube number. What do hex numbers have to do with cubes? For me these mathematical surprises connecting seemingly unrelated bits of the mathematical world are part of the magic of numbers. There is a connection because a hexagonal number is exactly the number of tiny cubes you need to wrap around the large cube to expand its size as the following picture illustrates:

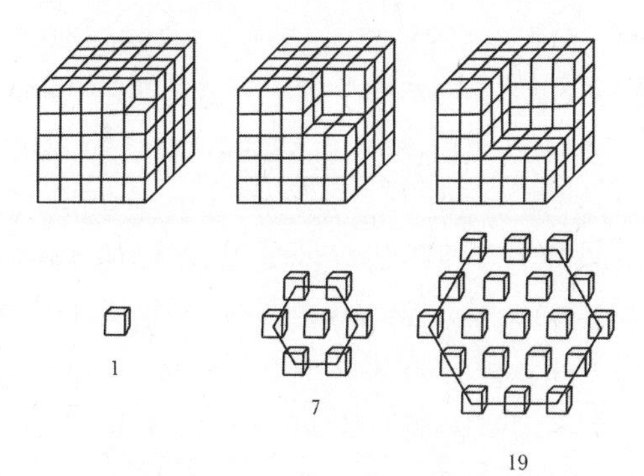

This is the trouble with being a mathematician: as soon as I first saw the beautiful arrangement of pieces on the carrom board, I couldn't stop heading off down a mathematical rabbit hole. And that's before anyone has flicked a piece across the board.

Once the carrom board's nineteen pieces are set up, each player has an extra circular piece, which is larger than the carrom men. This is the striker, like the cue ball in snooker, which players flick across the board to try to get the carrom men into the four corner pockets. Launching the striker is done with a flick of the fingers within two boundary lines marked out on the board in front of the player. You can place the striker at any point along these two lines in order to get the best angle to strike the carrom men.

The game proceeds much like a game of pool, with an interesting twist. The queen piece is a bit like the black eight ball in that it plays a special role. In carrom, after potting the queen, you must immediately pot one of your pieces on top of it. This is called "covering the queen." If this is done, then the queen is claimed and scores you five points. If you fail to cover the queen, it comes back onto the board.

As soon as I saw the game, I knew I had to take a carrom board home. So I trotted down to the market where the boards were sold. Along with the board and pieces, the other important ingredient is the powder that is spread on the board to ensure the pieces glide smoothly.

Originally this was boric acid powder. However, since the powder can cause infertility or damage an unborn child, it has now been put on the EU's list of substances with a serious health hazard. I decided to leave the boric acid and instead resorted to something called "disco" powder recommended by the UK carrom board association.

I ended up paying for extra baggage to get the monster of a board home. But it was worth it. My family has enjoyed endless hours lining up the mathematical angles to flick the carrom men into the pockets.

Although carrom has not caught on globally like its compatriot chess, another game with its origins in India has become universally popular, played everywhere from the cafés of the Middle East to the bars of South America to the bedrooms of children around the world.

13: Ludo

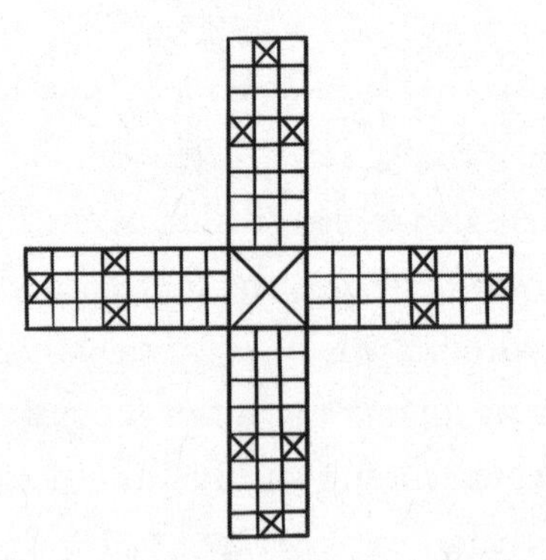

IF YOU VISIT THE PALACE IN Fatehpur Sikri, a town not too far from the Taj Mahal in Agra, you'll find a huge gaming board laid out in the courtyard. It is used for the Indian game of pachisi, the ancestor of modern-day Ludo, as I know the game, although it has a range of other names across the world. The sixteenth-century Mughal emperor Akbar so loved the game that he had huge boards built into the courtyards of the palaces he commissioned.

He would sit with his opponents in the middle of the board on a raised platform, casting the cowrie shells used as dice, and sixteen young women from the harem dressed in the players' colors were moved around the board as human counters.

Akbar was an enlightened emperor who preferred to persuade the locals rather than control them by force. He was tolerant of his non-Muslim citizens and scrapped the taxes they had been required to pay. Culture boomed under his rule, including the playing of pachisi, which underwent something of a rebirth during this period. The playing of games was encouraged, serving as a political tool to create a content and loyal population.

The game had a long history in India. There is evidence of it being played as far back as the sixth century, and some believe that a game of pachisi or its more aristocratic version, chaupur, was played in the ancient story of the *Mahabharata*. In the story the Pandavas lose everything over a game to their cousin Shakuni. It is said that Shakuni used loaded dice in order to trick the Pandavas out of all their possessions. The game of chaupur is played with dice rather than the cowrie shells used in pachisi, which is why some have speculated that it is this version of the game that is used in the *Mahabharata*.

The game is somewhat unusual in being a four-player board game. Up to this point in history, most games pitched two players against one another across a board. This made sense for games that simulated war, because it was natural to pitch one adversary against another. Yet there are many games that really don't work with just two players.

Many card games rely on a player being unsure where the cards they don't hold might be. With more players, the uncertainty, and therefore the enjoyment, increases. Modern board game designers strive to create games where the number of players can vary and the game still works.

But invariably these new games are at their best when played with three or four rather than two players. Pachisi's great advancement was in revealing that racing games needn't be limited to two players, and since the game was being used as a tool for bonding a group, the more players it could involve, the more effective it was at achieving its ultimate aim.

In pachisi, each of the four players occupies a different corner of the board. As in Ludo, the aim of the game is to race your four pieces around the board and into home. Certain throws allow you to introduce your pieces into the fray. In pachisi, the rules of the game mean that often players work in teams of two. The excitement comes from the fact that by landing on a square occupied by an opponent, you can send the opponent's piece back to the beginning. It shares a very clear ancestry with the racing games of the ancient Middle East, although it is unclear whether the game evolved independently or grew out of encounters with these early racing games.

A slightly unusual characteristic of the game is captured by the fact that the word *pachisi* means "twenty-five." The moves in the game are dictated by the throw of cowrie shells, which can land with their opening face up or down. In games of pachisi, six or seven cowrie shells are used. But rather than pieces simply being moved the number of squares indicated by the number of face-up shells, a slightly different set of rules apply. If six cowrie shells are used, two or more mouths facing up are counted as the number of moves you make around the board. But if only one mouth faces up, you get to move ten steps. If no cowries land face up, you get to move twenty-five squares—hence the name. Pieces enter the fray if you throw zero, one, or six shells face up.

Certain squares on the board marked with crosses are considered safehouses, like castles that you can hide inside to avoid capture. The castles at the end of each arm are particularly important because they are exactly twenty-five steps away from getting a piece home along the next arm. One strategy is to try to get your piece into the castle on the third arm from your home and wait to throw six cowries with no mouths up so you can get the piece home in one go.

Because you are playing with a partner, it is no good rushing your pieces home if your partner is proving slow in getting around the board. If you finish too early, then your opponents throw twice for every one throw your partner makes. There is the option to skip heading home and go round the board twice. In this way you might help capture your opponent's pieces, sending them back to base and aiding your partner in the race to get home.

The Indian game of pachisi started to go global as a result of the English presence in India. The families posted to India watched how fun pachisi was and soon brought it back to England, where it caught the imagination of Victorian society. A patent was received on August 29, 1891, for a new game that Alfred Collier claimed he'd invented. He called it Royal Ludo, *ludo* being Latin for "I play." But it is essentially a simplified version of pachisi. There are far fewer squares, making a faster game that is ideal for children. A similar game had already been copyrighted by John Hamilton in 1867 in the United States, where it is still known by the name that shows its connection with the ancient Indian game: Parcheesi. It was the best-selling board game in the United States until Monopoly came on the scene.

I have been struck by how many times in my travels around the world I have come across men sitting around a board in a café playing a version of this game that originated in India. I guess because it is regarded as a kids' game in the United States and the United Kingdom, I had somewhat underestimated its power as a game that adults might choose to play. But watching the games evolve in the cafés around the world, I can see why it is so popular. It sits in that perfect position where a game can turn on a single lucky dice roll. A player who seems to be in the lead suddenly has a piece sent back to the starting grid, and the balance of play shifts. It is a game full of drama and ups and downs, yet with simple rules that anyone can pick up in a few minutes. It is also a game that allows more than two to join, making it a perfect for those sitting around a café table. Perhaps it's no wonder then that Emperor Akbar chose it as the board game he had laid out permanently in the courtyards of his palaces, with members of the harem on hand to animate the game.

The Victorians also brought back another Indian game that has gone on to global recognition. While, like pachisi, it was once again adapted for children, its deeper meaning remains surprisingly complex and cosmic.

14: Snakes and Ladders

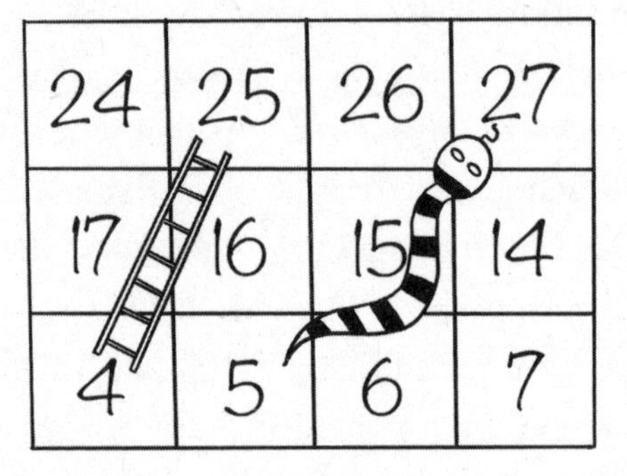

I WAS VERY SURPRISED to learn that Chutes and Ladders—one of my favorite games as a young kid, although I knew it as snakes and ladders—has its origins in India, and I was all the more stunned that it was in fact a serious tool in the spiritual quest for liberation from the vicissitudes of karma or the hindrances of passion. The first known boards date to the first half of the eighteenth century, but it is believed that they were developed centuries before. Versions can be found in Jain, Hindu, and Sufi Muslim traditions. But it is believed that its origins lie in Jainism.

This is a game emphasizing destiny, or dharma, which is the notion that your life is already mapped out for you, the only question being how closely you follow the path. In this case your destiny is controlled by the throw of a die or cowrie shells. In contrast with games where you might assert your free will in making decisions and crafting strategies, like chess, your trajectory through the game is completely out of your hands.

In India the game is known as Moksha Patam, where *moksha* is Sanskrit for the liberation of the soul from continuing rebirth. This is the aim in Hindu philosophy, to reach a state where you are no longer reborn to live another life. The snakes and ladders of the game give one the feeling of going round and round the board until finally you find your way through to the final square representing liberation. Your karma during life is represented by good deeds that take you up the ladders closer to *moksha* and bad deeds that lead you further away from this goal for the soul. With each square inscribed with its own different quality, the board was probably developed more as a document expounding the connections between karmic cause and effect than as a game, at least at first. The use of ladders as an image for the purification of the soul is already in Jain texts before the game was developed.

The Victorians who brought versions of the game back to England from India likely appreciated the way the game represented the impact of good and evil. A lesson in the morality of one's actions taught through the game was thought to be worthy for young children—although this take on the game rather misses the point that actually you don't really have any control over whether you are punished with a snake or rewarded with a ladder. In practice, the game probably taught Victorian children a lesson that moralists never intended: that it doesn't matter what you do; fate simply has it in for some people.

In the Indian versions of the game, the number of snakes outweighs the number of ladders: often nine snakes to four ladders. This is meant as a message to players that *moksha* is hard to achieve. The English version was more forgiving, with an equal number of snakes and ladders. Most modern versions of the game have removed the moral lessons that the game entailed: the ladders originally represented virtues such as generosity, faith, and humility, while the snakes represented vices such as lust, anger, murder, and theft.

One of my favorite museums in Oxford, the Pitts River Museum, has a rather beautiful board from the Punjab or northern Rajasthan, dating to the nineteenth century. Every square has a different quality described both in Hindu and Muslim terms, allowing it to be used by members of either faith. For example, square 54 corresponds to *bhakti*, or devotion, while square 7, at the bottom of a snake, is *mada*, or drunkenness, which you end up on as a result of landing at the top of the snake on square 24, which is associated with bad company, or *kusang*. As is typical of many similar boards in India, this one is laid out on a 9 × 8 grid of seventy-two squares. An old museum label attached to the board rather grandly refers to it as a "philosophical game board." The winning square is actually at the center of the top row on square 68 and is presided over by a four-armed Ganesh, the elephant god, who sits in a domed pavilion. If you reach square 72 you have to be reborn and head back to the beginning of the board again.

The mathematics behind reaching *moksha* are more complex than they may seem. In fact, the game of snakes and ladders is a perfect vehicle for explaining an interesting piece of mathematics that is highly useful for similar chance-based board games: Markov chains.

The Math of *Moksha*

The key to analyzing the inner workings of this game is something called the *transition matrix*. This is a square array with a row corresponding to each possible square you might be on. In the case of the board in the Pitts River Museum there are seventy-two squares plus a square that corresponds to your starting position before you enter the board. Let's call this square 0. The actual number of squares is smaller than this because you never occupy the head of a snake or the base of a ladder. These positions move you to different squares. There are ten snakes and ten ladders on the Pitts River board. So the square array just consists of 53 = 73 – 20 rows.

To fill the square array in each row we insert in the column the probabilities of moving to a new square corresponding to that column number after throwing the dice. So, for example, in the first row corresponding to being on square 0, the probability of being on square 1 is $\frac{1}{6}$ because there is a $\frac{1}{6}$ chance of throwing a 1. Similarly for squares 2 to 6. So the first row of the transition matrix looks as follows:

$$\left(0,\frac{1}{6},\frac{1}{6},\frac{1}{6},\frac{1}{6},\frac{1}{6},\frac{1}{6},0,0,\ldots 0\right).$$

Most of the rows actually look like this with the numbers gradually shifting to the right. The exceptions are where there is a chance of landing on a snake or a ladder. For example, if you are on square 9, then a throw of one will land you on a ladder taking you to square 23, while a throw of three will land you on a snake taking you to square 8. So the row corresponding to square 9 will actually look like this:

$$\left(0,0,0,0,0,0,0,0,\frac{1}{6},0,\frac{1}{6},\frac{1}{6},\frac{1}{6},\frac{1}{6},0,0,0,\ldots,\frac{1}{6},\ldots 0\right)$$

where the last $\frac{1}{6}$ is in the column corresponding to square 23.

(Remember that there is no column corresponding to the bottom of the ladder on square 10 or to the head of the snake on square 12, which is why there aren't zeros in the middle string of $\frac{1}{6}$s.)

Once one has built this matrix, the wonderful thing is that the mathematics kicks in to help you calculate efficiently the expected number of turns that it will take to make it to the winning square 68 and achieve *moksha*. We begin by deleting the row and column corresponding to the winning square of the transition matrix to get a 52 × 52 square matrix. We then subtract this matrix from the 52 × 52 identity matrix with ones down the diagonal. Then we apply an operation called *inverting the matrix*. The first row of this new matrix then has the expected number of times you will be on each square of the board.

Add up all the numbers in the first row and—bingo—that is the number of turns it takes in general to reach *moksha*.

When I ran the numbers for the board in the Pitts River Museum on my computer (I can't invert 52 × 52 matrices by pen and paper!), the result was that the expected number of turns to reach *moksha* was fifty-nine. That is more than in a Western snakes and ladders game in which you are trying to reach the last square and you have to throw an exact number to end. So you can spend a little time waiting. However, in the Jain version of the game, you get reborn and sent back to the beginning, which means that you are going to be going round the board for longer. The Western version on a 10 × 10 board with ten snakes and nine ladders has an expected thirty-nine moves to get to the end.

This matrix also allows you to test the effect of adding more snakes or ladders to the board. In general, of course, more ladders should shorten the time on the board, while snakes should lengthen it. But not always. It turns out that sometimes an extra snake can shorten the game. Why? Because the snake might take you to a point where you get a second chance at climbing a really long ladder that you might have missed on your first pass through the board. For example, when I removed the snake going from square 52 to square 35, the number of turns to finish the game shot up to seventy-seven. This snake actually helps you to reach *moksha* and win the game because it gives you another chance at climbing the ladder from square 37 to square 66. Perhaps there is a deeper, enlightened meaning to be gleaned from this: the game may have been intended to discuss the difficulty of avoiding bad karmic actions and even the possibility that some of those bad actions are, in some way, necessary to achieving the final goal.

India is not only the birthplace, as we've seen, of the greatest war game played on the planet but also the source of one of the landmark multiplayer games, as well as the home of a surprisingly deep game that captures a fairly profound religious truth. The Indians also may have innovated a huge revolution in game playing that took the game off a board and allowed you to carry it with you in your pocket: cards.

15: Ganjifa Cards

THOUGH MANY OF THE GAMES we've encountered till now are board games, card games represent an important slice of the games humanity has developed. The advantage of cards is that they are fantastically portable, and the same pack can be used to play a huge range of games. I rarely pack a board game on my travels, but I'll always take a pack of cards with me.

Despite the huge amount of research that has gone into tracing their ancestry, the precise origin of card games remains somewhat opaque. But the cards I discovered that were popular in India are certainly among the earliest examples of cards made for gaming. Called Ganjifa cards and used for trick-taking games, they originated in the Mughal courts and became popular in India under the Mughal emperors in the sixteenth century. Curiously, I first encountered these cards not in India but in London.

The Victoria and Albert Museum has a beautiful collection of these Indian cards. Although the Victoria and Albert doesn't have any particularly old sets, the Topkapi Palace Museum in Istanbul has a set of related cards that dates back to around 1500. These cards are actually from the Mamluk sultanate around Egypt, further muddying the waters as to the cards' origins. There are remnants of cards that are even older, perhaps created in the thirteenth century.

The first thing that immediately drew my attention to the Ganjifa cards was their shape. They are very often circular. They are beautifully hand painted, especially the court cards. There are also a lot of cards in a single pack. Although they might look very different from our European packs, the structure of the pack reveals that they are deeply connected. A Ganjifa pack most often has ninety-six cards, divided into eight suits. In each suit you have cards numbered from one to ten and then two court cards: the *shah* or *mir*, which is the king, and a lower court card called the *wazir*, which corresponds to the minister or advisor to the king.

There are some important resonances with chess here because the early versions of the board game had a similar pairing of king and *wazir*. Could the pack of cards have been created as a portable version of chess? Later packs of Ganjifa cards often see an additional court card added, which ultimately tallies with the packs used in Europe. There are no such court cards in China's decks, which are meant to have been the source of all other cards. So it could well be the meeting of Chinese cards and chess that inspired the addition of court cards.

The court cards provided the painters of the cards a canvas on which they could show off their skills. The king is most often depicted on a throne, attended by servants either shading him with a canopy or fanning him with peacock feathers. The king will sometimes hold a banner with the symbol for the suit he represents, or sometimes one of the attending courtiers is handing it to him. The *wazir* most often is shown riding an animal in his card—a horse or sometimes an elephant, camel, or tiger.

The suits correspond to a range of different objects. Perhaps the most significant ones are the two suits *safed* and *surkh*, which correspond to a silver and a gold coin, respectively. The word *ganj* is Persian for treasure, and the cards always have these two suits dedicated to coins. This constant in the cards might give a hint as to the original source of card games in general. The fact that coins are always part of the sets that one finds in India and Asia more broadly hints at a connection with the money cards of China. The circular shape of the Indian cards might also be a nod to the cards in China originally representing coins. The creation of cards requires the ability to manufacture paper, which has its origins in China. There is evidence of the Chinese using money cards before the year 1000, but the games don't appear to be related to those emerging in the later centuries. There is some argument that although cards might have started in China, there was a significant evolution over the centuries to create cards with the suits we see and the trick-taking games that are played with them. Perhaps these Chinese cards inspired the Ganjifa cards that would emerge in the Indian subcontinent.

The other suits vary quite a bit across different regions. In Ganjifa cards from the Mughals, we see crowns (*taj*), swords (*shamsher*), servants (*ghulam*), harps (*chang*), documents (*barat*), and textiles (*qimash*). Most of these suits are depicted quite simply except for the servants, which might often see the illustrator having fun creating scenes with ten attendants crammed onto the card. Colors would be used to help players distinguish between suits. For example, the harps would come in dark or olive green, while swords were crimson.

In later packs in India, we see these suits becoming Hinduized as they transform into representations of incarnations of the god Vishnu. The suits also enlarge in number to match the ten different incarnations.

There is some evidence to support the fact that originally packs just consisted of four suits totaling forty-eight cards. But packs were doubled up to make games more challenging. This is why we often see two suits corresponding to coins in one pack. Since these cards are typically used in trick-taking games, the ability to remember what cards have already been played is key to playing well. With more cards in play, this skill is pushed to its limits.

An intriguing rule of the games played with these cards relates to who opens the play. In European games it's usually the person to the left of the dealer. But in games of Ganjifa, the person who holds a particular card opens play. I love that the card changes according to whether the game is being played at night or in the day. This rule has parallels in the African game of mancala, where playing at night or during the day carries similar importance.

Another curious feature of the game is that unlike trick-based card games in Europe that hinge on the strategic use of high and low cards at different junctures of the game, the Indian rules dictate when high cards must be played. It is a strange feature because it lessens the role of strategy in play and places more emphasis on luck. As in other Indian games we have discussed, there is the sense of a person's destiny being determined and not in their control. But the game also places an important emphasis on memory. If you don't keep track of the fact that you have the most powerful card in a suit and fail to play it, then the card gets burned and loses its power to win a trick. The pack was probably doubled in order to increase the likelihood of cards getting burned.

The importance of a good memory plays a role in many card games. Remembering that the ace of spades has already been played in whist or bridge means knowing that the king can't be beaten. In rummy, if you are collecting queens, then you need to remember if any have already been discarded. In the casino, keeping track in blackjack of how many high cards have been dealt from the pack helps with assessing the odds of going bust or not. Card counting was developed as a technique to assist the player in keeping track of the cards without memorizing them all as they are dealt. Much like Ganjifa, casinos now put together multiple packs at the blackjack table to disrupt the ability to predict the cards left in the undealt stack.

Another peculiarity of the two packs put together in Ganjifa is that in half the suits the cards increase in value as you'd expect, from one to ten, with ten being the strongest card. But in the other half this is reversed, so that ten is the weakest card. It is as if these cards are numbered with negative numbers. In mathematics –3 is bigger than –10. And just as the number zero originated in India, there is evidence of the mathematics of negative numbers being developed alongside the investigation of the properties of zero. In the seventh century, for example, Brahmagupta investigated the result of multiplying two negative numbers together. The result, he concluded, should be positive.

Though some have argued that European cards developed independently of the cards of Asia, there are simply too many coincidences supporting their Asian origins. One in particular relates to the ordering of the cards. In Spain and Italy up to the fifteenth century, two of the four suits used in games were ranked in reverse order like the Ganjifa cards. It seems very likely that the trading Spain and Italy did with Asia involved not just spices but also games. They also traded mathematics, explaining the arrival in Europe of Hindu-Arabic numerals. The use of black and red in cards might connect with the Chinese use of these colors to indicate positive and negative numbers. Although European game players seemed happy to use these negatively ordered cards in their games, Europeans took a long time to be happy with the abstract concept of negative numbers. Both zero and negative numbers were banned in some Italian states as numbers of the devil.

I must admit that on my trips to India I never actually witnessed Ganjifa cards being dealt at the tables of cafés across the country. Mass-produced European cards have flooded the market and pushed out the artisans who once would have painted you a set of circular cards. Today just a few studios make the cards in the eastern state of Odisha in India.

Over the course of its long and rich history, India has created some of the most important games that have captured the global imagination. The passion for mathematics and games on display in the country makes it one of the places I enjoy going back to again and again. But not everyone agrees with me about the value of India's games: one of its local celebrities, who became one of the most renowned and influential men who ever lived, didn't seem to enjoy playing games at all.

16: The Buddha's Banned Games

THOUGH SOME PHILOSOPHERS OF reincarnation found games ideal teaching tools, the Buddha disagreed. For the Buddha, games were a distraction from the serious task of reaching enlightenment. In his discourse *Brahmajala Sutta*, written in the fifth century BCE, he lists the games he believes his followers should avoid at all costs. It provides a fascinating insight into the games that were being played at the time.

Here are the games banned by Buddha:

1. Games on boards with eight or ten rows, which probably refers to forerunners of chess.
2. The same games when played on imaginary boards. Playing chess in your head as you meditate is just as forbidden as the real thing.
3. Games where lines are chalked on the floor and you jump between the lines. This seems to refer to an early version of hopscotch.
4. Games where pieces are added or removed from a stack where the loser is the one who topples the stack. Jenga, therefore, was also banned.
5. Games that involve hitting big sticks with small sticks.
6. Anything with dice.
7. A kind of party game where you dip your hand in red dye and strike the floor or wall, calling out, "What shall it be?" Other players are tasked with identifying an animal in the handprint. Thus an early form of Pictionary was out.
8. Anything involving a ball.
9. Blowing through a *pat-kulal*, a toy pipe made of leaves.
10. Somersaults.
11. Playing with toy ploughs, windmills, carts, or bows.
12. Guessing words traced out on your back by an opponent.
13. Guessing your opponent's thoughts. Twenty questions was thus ruled out.
14. Imitating deformities, something that we might struggle to even understand as a game.

Reading, as it does, like an entry in a Borges story, I'm not sure if this list leaves a single game that the Buddha would have been happy to play. Still, his loss is our journey's gain, since he points the way to a few more games with origins in India that we may explore.

17: Hopscotch

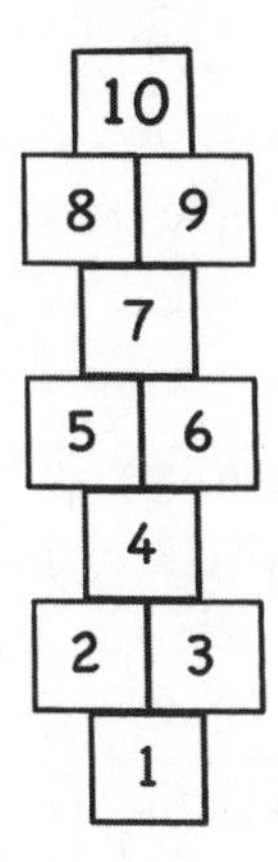

IT'S INTERESTING THAT THE BUDDHA chose to ban hopscotch because the origins of this street game turn out to be more meditative than I would have expected. It appears that the journey from the first square through to the last is once again a representation of a spiritual path to paradise, reminiscent of the idea of snakes and ladders.

A reminder of how the game is played: You chalk out on the ground a numbered grid of squares. At each turn you throw your stone into successive squares, then hop down the grid and back, avoiding jumping in the square with the stone inside.

The game gets an explicit mention in 1677 in the *Poor Robin's Almanac*, a satirical journal in England penned by an invented character called Poor Robin of Saffron Walden. So popular was the almanac that it ran from 1663 to the Georgian period. Poor Robin was described as an author, astrologer, journalist, satirist, and "well-willer to the mathematics." The almanac is full of mathematical curios and strategies for the reader to explore.

The scotch in hopscotch has nothing to do with Scotland but refers to the scratching of a line in the ground. An 1886 paper presented to the Royal Anthropological Institute of Great Britain and Ireland by a J. W. Crombie documents the many different designs across Europe that were used to play the game before the modern-day grid that you'll see drawn out in the playground. Many of these designs hint at early connections with Christian scriptures.

The early designs were often made up of seven regions, a number with a deep cosmic resonance. Seven corresponded to the number of transient heavenly bodies that the ancients could see with the naked eye—the sun, the moon, Venus, Jupiter, Mars, Mercury, and Saturn—which is why it assumed cosmic significance.

This led to many religions dividing heaven into seven regions, hence the phrase *seventh heaven*. So the hopscotch grid might represent this division but also might simply represent a pilgrim's path to paradise.

The final square in the grid was often marked with the word *paradise* or *heaven*. This is where one was aspiring to reach. But on the way you might have to navigate hell, purgatory, or limbo before attaining salvation. The layout was meant to mimic the floor plan of a church or basilica with the altar at the end of the grid. This is perhaps why you start to see a cross-like figure emerging where you could land with both feet rather than continuing to hop.

The meditative nature of the hopscotch court has some resonance with the idea of the labyrinth that one can find in Chartres Cathedral. This labyrinth, not a maze, only has one path taking you back and forth until you arrive at its center. It is used by visitors to the cathedral as an aid to prayer.

Curiously the original name that Buddha uses to refer to the game he banned can mean a circle with lines on the ground that you should avoid stepping on. French versions of the game refer to it as *escargot*, or "snail," indicating a similar spiral structure.

A common myth about hopscotch claims that it was invented by the Romans and used to train their soldiers. Decked in full military armor, soldiers were required to hop their way through the grids marked on the ground in order to improve their footwork. But the self-styled rogue classicist David Meadows believes this myth is the result of a confused reading of a paper published in 1870 in the *Journal of the British Archaeological Association*, in which an archeologist suggests that some of the pottery discovered in a site in Warrington "would be admirably fitted for our modern game of hop-scotch."

During the first lockdown of the coronavirus pandemic in 2020, Jenny Elliott, an urban designer and landscape architect based in Edinburgh, started a hopscotch grid and then left her piece of chalk and a handwritten sign with the request "Please take a chalk... for... HOPSCOTCH! Add some squares to the hopscotch on your walk up the hill, and see if we can make it to the top (Bruntsfield Place) before it rains."

Amazingly for Edinburgh in spring, it didn't rain for another two weeks. By the time the rain washed the chalk away, the residents of Edinburgh had gone way past Bruntsfield Place and had added fourteen hundred squares. The only problem was that the chalk kept running out. At a time when many people felt overwhelmed by a situation that was out of their hands, the activity was empowering. It was a way of taking control of the local environment in a tumultuous and frightening time—and what better means of doing so than a surprisingly ancient game with a cosmic message of its own? It may have been on the Buddha's banned games list, but he surely would have approved of the sentiment.

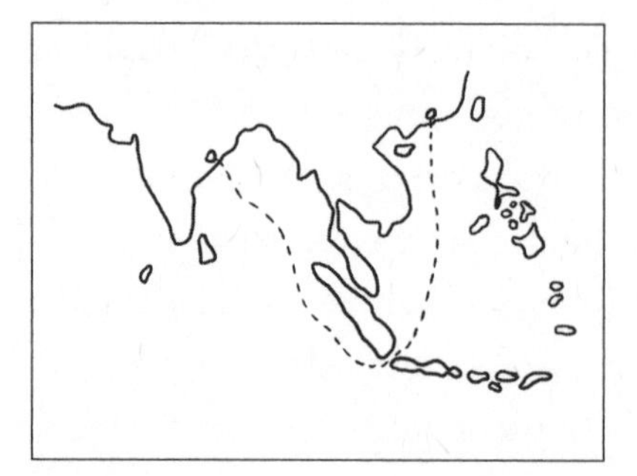

CHAPTER 4

THE SOUTH CHINA SEA

The Math of Games and Games of Math

AS WE PASS FROM INDIA and its numerous religiously inspired games (and one particularly notable war simulation, chess), we will use our sojourn at sea to shift gears and prepare for a set of games that reflect a fascinatingly different approach to culture, including another strategy game that couldn't be more unlike chess. In such games, mathematics plays an increasingly vital role. So in our passage from India to China, I want to take the opportunity to explain how mathematics is such an important part of making, playing, and, more importantly, winning games.

Why is it so important for us to win when we are playing a game? Simple: we want that dopamine hit that our brains get when we achieve checkmate, roll a six to win, make that hand of six no trumps, or whatever the relevant operation may be. One reason we engage in playing is because we crave the rush of drugs our brains reward us with when we win.

But getting that hit we desire requires a certain element of danger. We have to know that we could lose for winning to carry any value. And yet the strange thing about playing games is that even if you do lose, it isn't such a big deal. You can persuade yourself that it isn't that important. Of course, if you win, you get that rush because of the investment you've made in playing the game. It is important to care about the game as you play. In a way, games therefore represent a strange non-zero-sum game where the winner benefits more than the losers lose. It's a kind of perpetual-motion machine for feeling good. No wonder we play.

Mathematics helps us understand why we keep playing even if we lose—after all, as we just saw, the value we derive from winning is greater than the sense of loss we get from losing—but it also helps us swing the odds in our favor. The pages that follow will reveal how mathematics is really the key to maximizing your dopamine hit. Knowing your math helps you win.

18: Chocolate Chili Roulette

ON MY TRAVELS AROUND THE WORLD, I often visit the local schools to talk about my passion for mathematics. There's a game I always enjoy playing with students that illustrates the power of a mathematical way of thinking about the world to give you an edge. It's called chocolate chili roulette.

The game opens with a pile of chocolates and a chili pepper on the table. On each turn, a player can decide to take one, two, or three chocolates. When there are no chocolates left, the player whose turn it is loses and has to bite the chili. The key to winning relies on knowing how many chocolates are in the pile. When I play the game, I use thirteen chocolates, and on the first round I arrange to go first. This is a game where going first and playing carefully guarantees you a win.

The opening strategy is to take one chocolate, reducing the pile to twelve chocolates. Then, instead of seeing one pile of twelve chocolates, you regard this as three piles of four chocolates. The point is that whatever move my opponent makes—taking one, two, or three chocolates—I can always respond by taking the remaining chocolates from the pile of four. For example, if my opponent is greedy and takes three chocolates, I respond by taking one, with the result that I have cleared one of the piles of four chocolates.

Proceeding in this way, I can always use my move to clear the next pile of four chocolates eventually leaving my opponent with the chili to eat. Having played one round and forced my opponent to eat the chili, I explain that I am using a secret mathematical strategy to play this game and that it gives me the edge if I start. On the second playthrough, I let my opponent start. The beauty of this game is that the moment my opponent makes a slip, I get the upper hand and am back in control.

For example, if my opponent starts by taking more than one chocolate at the beginning, they've made the mistake of starting to eat into the first pile of four chocolates, so I can immediately respond by clearing the remains of that pile and putting myself back in a winning position. Or suppose they take one chocolate; I respond by taking one chocolate from the first pile of four. If they fail to clear the pile, taking, for example, just one chocolate, then I'm back in because I can take the two remaining chocolates.

It is fascinating to see the gears of my opponent's mind at work as they try to glean what it is I'm doing each time that forces them to eat the chili even when they start. I remember one occasion when I was in something of a rush to get to the school and had failed to pick up chilies in advance. At the station on my way, I dove into a local grocer to get some chilies.

When I got to the school, I was already rather late, so we went straight into the session on using math to win at games. When we got to chocolate chili roulette, lots of hands went up to play the game. Chocolate is a great incentive for getting volunteers. The girl I chose came to the front and lost the opening game. She bravely bit the chili and chewed. I could see that she was having trouble with the heat. But she was looking forward to beating me, so we played another round, this time with her starting.

She slipped up, and I was able to get the upper hand again to make her eat another chili. This time I could start to see sweat forming on her brow as she munched. I offered her the chance to sit down and let another student try, but she was determined to crack the game. We played again. She slipped up and lost again. She ate another chili. At this point I could see she was physically shaking from the heat. I decided to save her from another bite of the chili and decided to explain the strategy I was using. In the last round I let her apply the algorithm to force me to eat the chili that time.

As I bit into the chili, my eyes immediately began to water. I chewed and tried to explain the algorithm again. But I couldn't speak. The chili was so hot, I was unable to continue the talk. We had to wrap up early. If that was the effect of just one chili, no wonder the student had been sweating and shaking. I spent a good bit of the journey back googling whether you can kill someone with chilies.

While earthquakes are measured on the Richter scale, chilies are rated on the Scoville spiciness scale, a grading of heat that goes from the lowly bell pepper (0) right up to the fearsomely named Carolina reaper (2.2 million). The scale measures the concentration of the molecule capsaicin, which activates the neurons associated with pain by sending a message to the brain similar to the one sent when the body touches something hot. The sweating is the first response as the body tries to cool itself. The next reaction is an attempt by the body to get rid of the thing causing the pain.

My googling led me to the story of two journalists who took up the challenge of a local restaurant in Hove in Sussex to try their XXX hot chili burger. They ended up in the hospital after eating burgers that were at the far end of the Scoville scale. I decided I needed to phone the school to check up on the student. They reported that she was still standing and had earned a grudging respect for the power of mathematics and chilies!

19: Nim

THE MATHEMATICS BEHIND CHOCOLATE chili roulette is pretty straightforward once you've understood what's going on. But there's a related game, called Nim, with a really beautiful bit of math that supplies a winning strategy. In this game there are now several piles of chocolates.

Once again, the winner is the person who clears all the chocolates, leaving the opponent to bite the chili. However, this time you can take as many chocolates as you want from one of the piles.

You don't want to leave your opponent with just one pile because then they can take the whole lot and leave you with the chili. The aim is to keep piles active until you can force your opponent to leave you with one pile to clear. So how do you do it?

The secret is to use binary numbers. This is a way of writing numbers in terms of powers of two. Every number can be broken down uniquely into a sum of powers of two: for example, $9 = 8 + 1 = 2^3 + 2^0$, or $15 = 8 + 4 + 2 + 1 = 2^3 + 2^2 + 2^1 + 2^0$. We write this decomposition using binary numbers. A 1 in the nth position (counting from right to left) means that there is a 2^{n-1} in the representation. So 1001 represents $2^3 + 2^0 = 9$, while 1111 represents $2^3 + 2^2 + 2^1 + 2^0 = 15$.

This is really just the same thing as our decimal notation for numbers, but instead of powers of ten, we are using powers of two. Binary numbers are particularly favored by computers as a way of recording numbers because there are just two states: one and zero; on and off.

So how do binary numbers help us play Nim? Suppose, for example, that the three piles consist of four, five, and six chocolates. The trick is to turn the number of chocolates in each pile into binary. So four is 100, five is 101, and six is 110. A strange rule for adding these numbers together will help you see if you are in a winning position or not. Add the numbers in columns, but use the rule that $1 + 1 = 0$. So

$$\begin{array}{c} 100 \\ 101 \\ \underline{110} \\ 111 \end{array}$$

Your strategy is to remove chocolates from one pile so that this sum changes into 000. This is always possible. For example, if I take the five pile and remove three chocolates, then there are two chocolates left. Two in binary is 010. Do the sum again, and we get 000:

$$\begin{array}{r} 100 \\ 010 \\ 110 \\ \hline 000 \end{array}$$

The great thing is that whatever your opponent does next, they have to change this total to something with some ones appearing. If there are any ones, they definitely haven't won the game yet. But then you always have a strategy to reset the sum to 000. At some point that will result in your genuinely having removed all the chocolates from the table, forcing your opponent to eat the chili.

The language of binary numbers translates this game into one that you can always win, even if the number of chocolates or the number of piles changes. If at the outset the sum is already a string of zeros, then make sure you offer to go second. Otherwise, you go first and make a move that sets the sum to zero.

Nim is an ancient game, and there is evidence of a related game played in China called Tsyan-shizi, which translates as "picking up stones." In this Chinese variation of Nim, there are two piles only, and the rule is that you can take stones from one pile, or you can take an equal number of stones from both piles—an intriguing twist. Winning Tsyan-shizi requires an important divergence from our strategies for Nim and chocolate chili roulette, as we shall see.

How to Win at Tsyan-shizi

Despite the games appearing to be related, we actually need to use a rather unexpected way of representing numbers to play Tsyan-shizi. Playing this ancient game involves exploiting some of nature's favorite numbers: the Fibonacci sequence.

We begin by considering positions that you don't really want to find yourself facing. If the number of chocolates consists of one chocolate in one pile and two in the other, then you are sunk.

You can't win. Anything you do leaves your opponent in a position to remove all the chocolates. Here is a table of losing positions:

Pile 1	1	3	4	6	8	9	11	12	14
Pile 2	2	5	7	10	13	15	18	20	23

The aim therefore is always to try to make a move that leaves the game in one of these states for your opponent. But what is the key to remembering these numbers? Instead of binary we use a rather unusual way of representing numbers. The Fibonacci numbers are a sequence of numbers in which each subsequent number is generated by adding the two previous ones. So, for instance,

1, 1, 2, 3, 5, 8, 13, 21...

The sequence is named after the Italian mathematician Fibonacci, who discovered their importance to the way things grow in nature. But like so many things, this sequence was actually known in India long before Fibonacci came on the scene. Indian musicians discovered the Fibonacci sequence was key to counting the number of rhythms you can make on a drum from long and short beats. But just as with Ludo, snakes and ladders, and the numerals we use, other cultures appropriated these discoveries and called them their own.

To play Tsyan-shizi we again use strings of zeros and ones, but instead of indicating powers of two, the presence of a one will indicate using that Fibonacci number in the sum. For example, 11000 means add up the fifth and fourth Fibonacci numbers, so 11000 = 5 + 3 = 8. And 10101 means add up the fifth, third, and first, so 10101 = 5 + 2 + 1 = 8. But eight is also the sixth Fibonacci number, so we can also write it as 100000. Unlike in the binary system, there isn't a unique way to represent a number in the Fibonacci system. However, if we insist that the representation not have adjacent ones and that the rightmost position be zero, then the representation is unique, and every number can be represented in this way. Let's call this the *canonical Fibonacci representation*. For the number eight, the canonical representation is 100000; 10101 fails because there is a one in the rightmost position, and 11000 fails because it has two consecutive ones.

Now put the losing positions in their canonical Fibonacci representation and we get the following table:

Pile 1	10	1000	1010	10010	100000	100010	101000	101010	1000010
Pile 2	100	10000	10100	100100	1000000	1000100	1010000	1010100	10000100

The second larger pile results from adding a zero to the canonical Fibonacci representation of the first pile. The first pile is all those canonical Fibonacci representations where the rightmost one is in an even position (counting from the right). The second pile is all those canonical Fibonacci representations where the rightmost one is in an odd position. The key here is to show why, if you are not facing one of these losing positions, you can always play in such a way as to force your opponent into one. And conversely, there is nothing you can do in a losing position that will turn the tables and put your opponent in a losing position.

Suppose that the piles contain ten and fifteen chocolates. Change these numbers into their canonical Fibonacci representation (10, 15) = (100100, 1000100). The 100100 is already in the right form to be in pile 2, so we just need to pull chocolates from the fifteen pile to reduce 1000100 to 10010, the corresponding pile 1 number.

The interesting thing about these losing positions is that every number appears somewhere and only once. So as you are assessing your position and planning your move, you need to decide for each pile whether the number of chocolates is a pile 1 number (the rightmost one is in an even position) or a pile 2 number (the

rightmost one is in an odd position). Then you will try to remove chocolates to put the two piles in a losing position.

In the case that the piles contain nine and twenty chocolates, because (9, 20) = (100010, 1010100) has 100010 in the right form to be in pile 1, we reduce the twenty to be 1000100, the corresponding pile 2 number. The only problem comes when one number is in a pile 1 form but the other number is too small to be put in a pile 2 form. For example (24, 32) = (10001000, 10101000). The twenty-four is in the right form to be a pile 1 number, but the thirty-two is too small to be made into its corresponding pile 2 number. Now we need to use the possibility of reducing both piles by the same amount. Reduce both by twelve and we get (12, 20) = (101010, 1010100).

It must be said that mentally translating things into Fibonacci sequences or binary and calculating the winning move accordingly can take some getting used to. The upside is that once you've mastered the mathematics, you can guarantee a win every time.

The three games I've considered are great examples of how turning a game into the right bit of math can give you a strategy for winning. But the application of mathematics to games can have far bigger implications than avoiding taking a bite of a hot pepper or winning a stack of chocolates. The power of mathematics to win at games has given rise to one of the most important fields in modern economics: game theory. Since its creation in 1969, the Nobel Prize for economics has been awarded to fifteen game theorists, including John Nash, whose life story was made famous by the film *A Beautiful Mind*.

Game theory provides us with a number of interesting games that crystallize many examples of human conflict. Playing them can be a very serious matter, especially when you consider that they might provide humanity with a way to overcome dangerous situations that can escalate into unnecessary warfare.

20: The Ultimatum Game

ON A RECENT VISIT TO THE mathematics department at the Hebrew University of Jerusalem, I was asked to give a talk in Haifa at a conference in honor of a faculty member who had recently died.

The conference agreed to arrange a car to take me on the two-hour journey to Haifa. The organizers asked if I wouldn't mind sharing the journey with another professor at the Hebrew University. Sure, I said. Who is it? Only the 2005 Nobel Prize winner for economics, Robert Aumann.

I went into a minor panic. Stuck in the car for two hours with such an academic celebrity. What would we talk about? I spent the next twenty-four hours before the trip reading up on the groundbreaking work Aumann had done to win the Nobel Prize. He was a specialist in game theory, especially (and rather controversially) the game of warfare. His ideas have formed the backbone of some of Israel's rather hard-core stances in the Middle Eastern political arena.

In his Nobel Prize lecture titled "War and Peace," he argued that simplistic peacemaking can cause war, while an arms race, credible war threats, and mutually assured destruction can reliably prevent war. Based on his mathematical analysis of game theory, he argued against withdrawal from Gaza in 2005 and against ceding land to the Arabs under threats. In the light of the use of his own mathematical analysis to justify and support the actions of hard-core right-wing political groups in Israel, a petition to deny him the Nobel Prize garnered one thousand signatures from academics around the world. Given my left-wing political leanings, if conversation turned to politics, the car ride might turn rather tense. Better to stick to the math of game theory.

I decided I needed to bone up on the basics of the games he was expert at playing: ultimatum games. These are fascinating because they involve a good slice of psychology. The basic ultimatum game goes something like this: Suppose the driver of the car from Jerusalem to Haifa offers us one thousand shekels to split between us—with a twist.

The driver tasks Aumann with proposing a division of the money between the two of us. I am then given the opportunity to accept or reject the proposal. If it's rejected, then neither of us gets any money, and it's handed back to the driver.

What's the strategy for each player to play this game? If Aumann fairly divides the money, then sure, I'll agree. But Aumann knows that frankly I should vote for any proposal that gets me any money; otherwise I get nothing. So why not offer me the minimum amount possible: 1 shekel for me, and Aumann pockets the remaining 999. Since the money is lost anyway if I reject the proposal, wouldn't I be content with getting at least one shekel?

This game brings into the mix what is called the *Nash equilibrium*. Named for John Nash, this state is defined as a set of positions taken by the two players so that no player can increase their own expected payoff by changing their strategy while the other players keep theirs unchanged. The point is that if my strategy is to accept anything rather than going home with nothing, then if Aumann offers me more than one shekel, he can continue to increase his own takeaway without worrying about my rejecting the offer until he has allocated me the absolute minimum.

But this doesn't take into account the possibility that money might not be everything in my assessment of this game. My resentment at being screwed over might outweigh the gain of one shekel, and so I might be happy to reject the offer just to ensure Aumann doesn't get away with such an unfair split. At what point am I prepared to take the insult of being given less and accept the proposal? This starts to depend on what a shekel means to me. If I am very poor, then even getting a few shekels might be worth the insult. If I'm rich the money may mean less than the sense of injustice I feel.

Research in Indonesia found that even when the value on offer amounted to several weeks' wages, participants balked at being offered anything less than 30 percent of the pot. The same threshold was found in experiments in Israel. In general, an offer of 20 percent of the pot will be rejected 50 percent of the time. The interesting thing is that if I didn't know how much money Aumann would get in the game, then I might actually be content with getting something rather than nothing. Ignorance is bliss. It's the injustice that I can't deal with.

It turns out that we get a big dopamine kick, comparable to having sex, when we turn down an unfair offer. That might be worth sacrificing a small monetary gain. An experiment in Pittsburgh in 2014 on pedestrians recruited outside bars between 9 p.m. and 3 a.m. found that drunk players were more likely to sacrifice any monetary gain in order to penalize anything less than a fifty-fifty split. Being intoxicated seemed to increase the value assigned to not being screwed over.

The proposer also gains a utility that isn't simply monetary. A study of some cultural groups found that the value of a good reputation in the group is more valuable than getting more money. Mongolian and Kazakh players would more often than not offer a fifty-fifty split as they wanted to appear fair in the eyes of the group. Experiments have also shown that when men and women play the game against one another, the perception of attractiveness seems to elicit a higher offer. Is the proposer playing a longer game? Does the proposer believe that the impact of a higher offer might have utility once the game is finished? In one case the woman gave the man everything. In answer to the question about her decision, she wrote, "I want at least one of us to get something."

This game reveals that you need to score not just the monetary value of a win but also other factors that are more difficult to quantify. There is fascinating speculation that the sweet spot for how much you need to offer your opponent in order to get the proposal accepted is connected to one of the most important numbers in mathematics: the golden ratio. This number represents what many regard as the perfect proportions in art and nature. A rectangle's sides are in the golden ratio if the ratio of the length of the long sides to the length of the short sides is the same as the ratio of the sum of the two lengths to the length of the long side. And this is how it comes into play in the ultimatum game.

It seems that I don't mind accepting an offer from you if the ratio of what I get compared to what you get is the same as the ratio of what you get compared to the whole sum on offer. This works out at a 61.8 to 38.2 percent split. Clearly you are getting more, but there is a sense that since you are sitting on the whole pot and dividing it up, we feel less screwed over if the proportions match those defining the golden ratio.

The Math of the Golden Ratio

People often encounter the idea of the golden ratio as expressing the perfect proportions in art, architecture, and even music, but here I want to give you the nitty-gritty of how to engage with it mathematically. A rectangle of dimensions one unit by x units is in the golden ratio if $\frac{(x+1)}{x}=x$. That means x is a solution to the quadratic equation $x^2 - x - 1 = 0$. The Babylonians came up with an algorithm for solving such equations. Assuming x is greater than one, this gives $x = \frac{1+\sqrt{5}}{2}$, which is approximately 1.618. This number is known as the golden ratio. Both the Fibonacci numbers and the golden ratio have proved useful allies in playing games. But these numbers are in fact intimately related. If you divide one Fibonacci number by the previous number in the sequence, then the result is a very good approximation of the golden ratio. The fraction gets closer and closer the further through the Fibonacci sequence you go.

As a hint as to why this is the case, let f_n denote the nth Fibonacci number and consider the fraction $f_{n+1}/f_n = (f_n + f_{n-1})/f_n = 1 + f_{n-1}/f_n$. If the Fibonacci fractions converge to a number x, then x satisfies $x = 1 + 1/x$. In other words, x is a solution to the equation $x^2 - x - 1 = 0$. But as we saw above, this is the equation for the golden ratio. So the number that the Fibonacci fractions converges to is actually the golden ratio.

The golden ratio might help resolve the ultimatum game, but I realized I would have to get to grips with another game if I was going to engage with Aumann on our car journey to Haifa. The analysis of this game was one of the things for which Aumann had been awarded the Nobel Prize.

21: The Prisoner's Dilemma

ONE OF THE THINGS THAT Aumann is particularly interested in is the idea of repeated games. As you play a game again and again, you tend to learn something about the way your opponent plays. How does your strategy change with that new knowledge, and how can we quantify that mathematically?

Aumann's work explores another classic scenario in game theory: the prisoner's dilemma. Suppose that the driver discovers when we arrive in Haifa that we've wrecked the back of his car with food and drink on the way. "Who made all this mess?" But before we have a chance to point the finger at each other, the driver offers us a deal.

If we both own up, then he'll charge us $100 each for the damage. However, if Aumann owns up and I blame him, then he alone needs to pay $400, and vice versa. If each of us blames the other, then we each have to pay $300.

The driver will allow us to discuss our strategy as much as we want, but when the time comes, we must seal our decisions in an envelope unseen by the other passenger and hand them to the driver. Of course the best scenario would be for us both to own up. That way the damage will cost us $100 each. The thing is, I realize that if Aumann accepts the blame in the sealed envelope and I double-cross him by pleading innocence, I get away with paying nothing. Even if Aumann decides also to double-cross me, I'm still better off claiming innocence: I'll pay $300 rather than the $400 I would pay if I admitted guilt and he claimed innocence. So personally it makes total sense in game theoretic terms to choose to double-cross. But *collectively* that's the worst scenario because we end up handing the driver $300 + $300 = $600 in total.

The trouble is that we inevitably prioritize the personal over the collective. Remember that the Nash equilibrium is the position that each player takes so that no change in their choice would give them a better outcome. Once Aumann makes his choice, then for me choosing to own up to the crime always costs me more than double-crossing my fellow passenger. For example, if he owns up and I own up, then I pay \$100, but if I double-cross, then I pay nothing. So "owning up" is unstable. The Nash equilibrium is for each of us to choose to double-cross.

However, repetition gives this story an interesting new wrinkle. What if we were doomed to repeat the journey from Jerusalem to Haifa endlessly in a version of *Groundhog Day*? Can we use the decisions we make in the future as bargaining chips for the decisions we make today? For example, if I were to threaten to use future days to punish Aumann for double-crossing me, would this be enough to change the dynamic of this game?

If I value the money I have in my hand today the same as I value money in the future, then this threat is enough to tip the balance. But most of the time we value money we have in hand *more* than the money we might have in the future. Consider the effect, for example, that inflation has on the value of our money. Suppose each day the money I hold devalues by half. I've told Aumann that if he double-crosses me, I'll swap my decision from owning up to double-crossing him from that point on. What should Aumann do? If he goes with owning up, then he'll end up paying

$$\$100+\frac{1}{2}\times\$100+\frac{1}{4}\times\$100\ldots=\$100\times\left(1+\frac{1}{2}+\frac{1}{4}+\ldots\right)=\$200.$$

But if he double-crosses me on the first day, although he will get away scot-free to start with, he will end up paying for it in the long run:

$$\frac{1}{2}\times\$300+\frac{1}{4}\times\$300+\ldots=\$300\times\left(\frac{1}{2}+\frac{1}{4}+\frac{1}{8}+\ldots\right)=\$300.$$

So he will realize now that it is much better to also agree to own up to the mess.

The critical number in this scenario is an inflation value of $\frac{1}{3}$. If money's value decreases by any factor less than $\frac{1}{3}$, then it isn't worth Aumann's owning up. But above $\frac{1}{3}$, the future cost tips the balance in favor of cooperation.

Unfortunately, we don't have infinitely many days to keep playing this game. Even if we had to take the journey a million times, we would always be back to square one on the last day because I know that nothing lies in the future to change the situation. Reverting back to playing the original prisoner's dilemma means that I will double-cross. Following that logic, if we step back to the penultimate day, I know that nothing will change what I do on the last day, so what I do that day doesn't matter. So again I will double-cross—and so forth. Backward induction means I will double-cross on all days!

However, if I don't know when the last day is and there's a chance we could always make another journey, I will be playing as if there are potentially infinite journeys and my decision today will affect tomorrow's potential game. The game clearly grows more complex as we think about how it evolves over time—even more so if we take another cue from actual human behavior and choose to punish each other for double-crosses and reward one another for cooperation.

In 1979 political scientist Robert Axelrod invited programmers to submit algorithms to play the prisoner's dilemma in an iterated version consisting of two hundred rounds per game. He wanted to see which of the countless arrays of strategies would triumph in a computerized version of the game. For example, how would just tossing a coin and behaving randomly in your choices fare against a more strategic approach. In the first "Olympic" computerized prisoner's dilemma games that Axelrod staged, the winning strategy turned out to be a simple tit-for-tat (TFT) approach. In TFT you own up to the crime until your opponent turns on you, and then you just do back to them whatever they've done to you in the previous round. If they keep double-crossing you, keep double-crossing them back until they learn that it isn't in their interest. As soon as they own up to the crime, reward them by also owning up in the next round.

Axelrod attributed the success of TFT to four qualities. It is *nice*: it starts off by collaborating with its opponent. (The top eight strategies in the competition were all nice, as it turned out.) However, it is *retaliatory*: once the opponent chose to spike you and betray the trust,

TFT responds with punishment. At the same time, it is *forgiving*: if the opponent changes their behavior to indicate a willingness to collaborate, then TFT forgives on the next round and returns to behaving nicely. Finally, it has the virtue of *clarity*: Axelrod believed that it was important to signal to the opponent what the rules of engagement were in order that the opponent could understand how best to exploit this knowledge strategically for a mutually beneficial engagement. A strategy using randomness fails to provide this sort of clarity.

Since that first competition, many different strategies have been proposed. It seems that the TFT was particularly successful in the context of the rules that Axelrod put in place for that first competition. Change the rules of engagement, and other strategies prevail.

One fascinating aspect of this research has been considering what strategy would emerge in an evolutionary context. Aren't genes, for example, meant to be selfish? Wouldn't evolution always favor the individual over the group? How would a nice strategy like TFT fare against a nasty strategy that always betrayed? One way to play an evolutionary game is to put a bunch of players into the pool, say fifty Nice and fifty Nasty, get them to play each other, and assign three points if you end up paying $100, one point if you end up paying $300, and no points if you pay $400. Then, at the end of the round, you rebalance the proportion of Nice and Nasty according to how many points each player scored. For example, if the Nice players scored twice as many points as the Nasty players, then in the second round you'd put twice as many Nice players into the mix as Nasty players.

Provided there are enough Nice players in the mix, in the end they tend to dominate because they can generally find enough other Nice players to play against and together earn more points than the Nasty players. If you started with ninety-nine Nasty players and just one Nice player, then the Nice player just doesn't have anyone to generate points with. But put a couple of Nice players into the mix, and evolution leads to the pool eventually being dominated by Nice players. For a game that at first sight seems to reward selfish, noncooperative behavior, this is a surprisingly comforting conclusion.

Having spent twenty-four hours studying Aumann's contribution to the effect of repeating games on these strategies, I felt I was ready for my car journey with the Nobel Prize–winning game theorist. But when the car came to pick us up, Aumann didn't show. I took the car on my own to the meeting in Haifa. I just made sure I was careful with the food and drink in the back since I had no one but myself to pin the blame on for any mess made. The twenty-four hours of research convinced me even more that mathematics has a powerful role to play in planning your strategies when playing games.

Most people regard games and mathematics as polar opposite pursuits. It's generally only nerds like me who regard escaping into a mathematical equation as a joyful activity. But increasingly we are seeing a mathematical approach to games becoming a mainstream part of the mathematical landscape. Indeed, ever since the seventeenth century, when Pierre de Fermat and Blaise Pascal started applying mathematics to the roll of a die, we have seen mathematics giving people an edge in playing games.

One major strand of mathematical involvement in games is the concept of "solving" a game. This means being able to predict the outcome of a game from any position, assuming the players are playing perfectly. The concept of a solved game really only applies to games of pure strategy since we can't predict the roll of a die in a game of chance. For example, our analysis of Nim and chocolate chili roulette "solved" these games because I showed you an algorithm of how to play them to win from certain positions.

We will encounter a number of solved games as we make our way around the world. Some of these solved games are merely analyzed by brute force, considering all the different possible games, often with the assistance of computers. For example, the strategy for playing tic-tac-toe to a draw is a result of analyzing all the different possible games. We can use a bit of symmetry to narrow down the possible cases—an X in the top right corner is the same as an X in any corner because you can rotate the game. Checkers or draughts is another game that it has been proved will end in a draw if players play optimally. The proof was achieved by a brute-force analysis requiring 10^{14} calculations conducted over an eighteen-year period with two hundred desktop computers. These proofs are boring, giving no real insight into the game.

More interesting are proofs that don't provide a way to play the game but still prove, for example, that the first player can always win. Take the game Chomp. It is similar in flavor (literally) to chocolate chili roulette. A bar of chocolate has a poisoned piece in the top left corner. Players take turns choosing a piece in the bar and eat all chunks below and to the right of the chosen piece. The aim is to leave the poisoned piece for your opponent. If the bar of chocolate is big enough, we don't know a strategy to win the game, but we can prove that one must exist that wins the game for the person who goes first.

The Math of Chomp

The way to prove that player 1 has a winning strategy in Chomp is to use something called a *stealing strategy* discovered by John Nash. It's important in this game that there are no ties. Someone must win the game. From every position either (1) there is a winning move, or (2) every move loses you the game, something that can be proved by backward induction on the number of squares. Suppose that player 1 does not have a winning opening move. That means that whatever move player 1 makes on his first turn, player 2 has a move that wins her the game from that point. But player 1 has a way to steal player 2's winning move. Player 1 simply removes the lower-right square from the bar and waits to see what player 2 responds with. But since player 2 must make a move that could have been an opening move, player 1 can steal this move as his opening gambit in any subsequent game and win the game himself. The curious aspect of this proof is that it shows player 1 can always win but doesn't show how!

Once a game has been solved, there is a case to be made that it has ceased to be a game at all. People love playing the game Connect 4, where you alternate dropping colored counters into a 6 × 7 grid, trying to get four counters in a row. But don't bother playing if you are up against a mathematician. We've found a guaranteed strategy to win this game provided we start the play, thanks to another brute-force exploration of the tree of possibilities. Even for the mathematician who has worked out this strategy, there is little fun in applying it: What pleasure will they get when they win other than a feeling of superiority over their opponent? Play requires uncertainty, and a game like Connect 4 only maintains its game quality if played by those who don't know their math.

This is one of the pitfalls of mathematics in the world of games: there is always the possibility that it might work too well, perfecting one's strategy to the point that the outcome ceases to be in doubt, and no one wants to play anymore. After all, in chess, if you see that checkmate is an inevitability, you resign. Even if one is sure of winning, it no longer counts as play. As Roger Caillois writes, "The pleasure of the game is inseparable from the risk of losing." Thank goodness then for games whose complexity is such that computers will be hard-pressed to ever completely solve them. One of the greatest of these will take center stage in our journey's next destination, China.

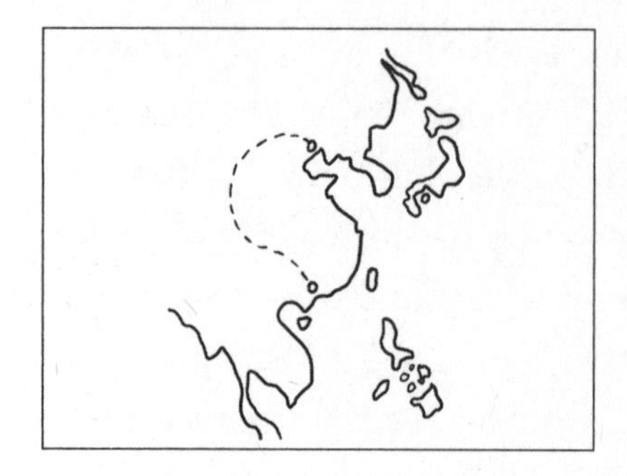

CHAPTER 5

CHINA

THOUGH I HAVE VISITED CHINA many times on my mathematical travels, it is probably the place that has felt most foreign to me of all the countries I have journeyed to. I have no Mandarin, and oftentimes my hosts had little English. And yet our mathematics gave us a common language with which to share stories. Their games too provided an interesting bridge. Although the outward appearance of many of their games felt quite foreign, once I understood the rules, it was striking how they too connected to things I recognized.

There is one game that originates in China that I believe does retain a very distinctive character from other games across the world. It can be viewed as a game of war like chess, where each player controls an opposing army of white and black forces. And yet the flavor of this encounter is very different from the combative narrative of chess. It is a game of territorial conquest, but rather than being removed from the board as if killed in combat, the pieces in this game build and grow an empire as the game proceeds. The pieces themselves are as simple as you can get: just black and white stones. The complexity comes from what you do with them.

22: Go

INDIA GAVE THE WORLD CHESS. China gave it Go. Both games mimic warfare waged by two armies, but the character of each game reflects the different culture from which it emerged. And players of Go are adamant about which is the more sophisticated. As the author Trevanian writes in his novel *Shibumi*, "Go is to chess what philosophy is to double-entry accounting."

Known in China as Wei-qi, the game is more commonly known in the rest of the world by its Japanese name: Go. It is regarded as the oldest board game in history to be played continuously to the current day. Some have tried to date the beginning of the game as far back as 2300 BCE, suggesting it was created by Emperor Yao to strengthen the mental faculties of his lazy son Tan-chu. Though the game provides a great workout for an unfit brain, most historians regard this as an unsubstantiated myth. Not until the sixth century BCE are there definitive records of a game like Go being played in the *Analects of Confucius*, so this is regarded as a more trustworthy origin point.

In contrast to the warlike quality of chess, Go is interested in claiming territory. Players take turns placing white and black pieces or stones onto a board marked out with a 19 × 19 grid. The grids are not perfect squares but are slightly elongated in the direction of the players to compensate for the foreshortening effect of perspective. The aim is to surround areas of territory with your stones. The player with the largest area of the board captured at the end of the game wins. Unlike in chess, where the pieces sit in the center of each square, the stones are placed on the points where the lines of the grid intersect.

Whereas chess depicts two armies engaged in a head-on battle, Go is about what happens when two armies arrive in an unpopulated land and try to settle more territory than their opponent. You may end up surrounding the opponent's pieces and capturing them, but the emphasis is on controlling the land, not defeating the opposing forces. It is rather unique as a game in that pieces don't move and are simply there to claim ownership of the land.

Nineteen is an interesting choice of number for the dimensions of the board. If you position yourself at the center of the board, you are left with $19 \times 19 - 1 = 361 - 1 = 360$ remaining intersection points. Given that the circle is divided up into 360 degrees, it is interesting to speculate whether this indicates that the board represents the entire lay of the land. However, throughout history Go has been played on different-sized boards, with 17×17 being popular in ancient times. The number nineteen seems to have emerged as the board size that gives rise to the most satisfying level of complexity while still being constrained enough for a game to end within a few hours' play.

It was found after some time that the player who takes the first turn has on average a five- to seven-stone advantage, so in the modern version of the game introduced in the 1930s, the second to play (who always plays white, in contrast to chess) is given a 6.5-stone bonus, called the *komi*, to compensate for the advantage black has by starting. The choice of 6.5 rather than 6 means that you never get a tied game.

A command of Go became one of the fundamental accomplishments that any Chinese gentleman should possess, along with the art of calligraphy, painting, and playing the lute. The Japanese took to the game some one thousand years ago, and by 1600 AD it was so established as part of the nation's life that four schools dedicated to playing the game were formed.

Representatives of these four schools would come together each year to battle it out in a series of matches called the Castle Games. The winner was made part of the government and appointed to the cabinet-level position of *go-doroko*, or minister of Go. During this period Go's sophisticated ranking system developed. There are nine grades, or *dans*, of professional player, the highest being awarded the title of ninth *dan*, a bit like being a black belt in judo.

Go playing is full of ritual. The etiquette of the game requires that players pick up and place the stones on the board between their second and third fingers. Although it is very tempting, running your fingers through the stones in your wooden bowl as you think about your next move is regarded as very bad form. The game is always preceded by the players bowing to each other and declaring *onegaishimasu*, which means "if you please."

Although the game appears simple, it has been recognized for many years as immensely more complex than chess. Given the size of the board and the freedom to place stones nearly anywhere, the number of possible games grows explosively. At any time in the game, each point can either be occupied by a white stone or a black stone or not yet claimed. Since there are 19 × 19 = 361 points you can play on the board, there are 3^{361} possible configurations. This is a number with 173 digits. It turns out that only about 1.2 percent of these positions are legal. In 2016 the number of legal positions was calculated as 208,168,199, 381,979,984,699,478,633,344,862,770,286,522,453,884,530,548, 425,639,456,820,927,419,612,738,015,378,525,648,451,698,519,643, 907,259,916,015,628,128,546,089,888,314,427,129,715,319,317,557, 736,620,397,247,064,840,935. This is a number with 171 digits. To put this in perspective, it is believed that you can count all the atoms in the observable universe with a number with only eighty digits.

So the number of positions on the board is large. But then you have to factor in that a game consists of a sequence of these legal positions. The American Go Association estimates that you need a number with three hundred digits to count the number of games that you can play. In chess, computer scientist Claude Shannon estimated that a number with 120 digits (now called the Shannon number) will suffice to count all possible games of chess.

Pattern recognition is a key part of being able to play Go well. Such a complex game does not so easily succumb to analysis that focuses on the effect of placing a stone several moves ahead like in chess. Rather, players build up a much more intuitive and less clearly articulated approach based on the patterns that begin to emerge. The other key difference from chess is that while chess gets simpler as the game goes on because pieces are removed from the board, Go increases in complexity because more pieces are added as the game is played.

This is one reason why it took many more years than for chess for a computer to be programmed to play Go at an expert level. It required the introduction of a completely new style of coding called machine learning for an algorithm to emerge that could play Go at anything like a professional level. Traditionally code was written in a very top-down manner. A human would write a set of instructions: "If this, then do that." And the computer would implement these at speed and depth. This was ideal for analyzing the game of chess and deciding optimal moves. But it didn't work for Go. Instead a bottom-up approach was needed where code could change and mutate as it encountered new scenarios in the game. This new method of writing code allowed the code to learn from its failures and to update itself, mutating and improving. This style of coding was much more suitable for challenges that involve pattern recognition, like developing image-recognition software.

But it also turned out to be ideal for creating code to play Go.

In 2016 the Go community witnessed a computer program playing the game at the very highest level for the first time. Called AlphaGo, the algorithm was able to beat the world's best Go player, Korean Lee Sedol, by four games to one in a match held in Seoul. Lee Sedol himself was shocked at the ability of code to play such a complex game so brilliantly. He knew that the eyes of the world were on him as their representative in humanity's battle against the machines. "I apologise for being unable to satisfy a lot of people's expectations. I kind of felt powerless." The one game he did manage to win he now regards as one of the most important of his career. "As time goes on, it will probably be very difficult to beat AI. But winning this one time, it felt like it was enough. One time was enough."

Three periods in the history of the game are celebrated as turning points in the way the game is played. The first of these occurred in the 1670s when Meijin Dosaku, the fourth head of the Haninbo School, introduced a whole new theory of the game. Dosaku is regarded as one of the greatest Go players in the history of the game, and over 150 of his games are recorded and analyzed to this day. It was said that his way of playing gave him a two-point advantage over his opponents. He introduced a new approach to openings and also placed an emphasis on *katachi*, or "good shape." He also developed interesting strategies including *amashi*, in which you let your opponent take what are regarded as the big points on the board, compensating for this by taking small pieces of territory on the edges with the aim of outlasting your opponent. It is something of an all-or-nothing strategy. In fact it was the strategy that Lee Sedol used in the only game he was able to win against AlphaGo.

The second Go revolution happened in the 1930s when Chinese-born Go player Wu Qingyuan, known as Go Seigen in Japan, developed a new approach to playing the opening of the game called *shinfuseki*. Traditional tactics emphasized strategies that went from corner to side to center as the order to attack territory. *Shinfuseki* challenged this, showing how playing in the center of the board can provide a powerful opening. As Wu Qingyuan explained, "The 'new opening' period was only a step on the way to the perfecting of Go... Go should be played over the whole board. In that respect Shinfuseki was an ideal style to stimulate the creativity of players beyond fixed josekis (corner patterns) and to broaden understanding of the game."

The clash of the old style versus the new is beautifully captured in one of my favorite novels, *The Master of Go* by Yasunari Kawabata, the first Japanese author to win the Nobel Prize. This is a semifictional account of an epic game that took place between the last head of the Haninbo School, Meijin Shusai, and the young upstart Kitani Minori, a contemporary of Wu Qingyuan, called Otaké in the novel. A strange little detail of the book that has stayed with me is the master's insistence that his barber leave one very long hair growing in his left eyebrow because it is regarded as a sign of good luck and longevity. The detail has stopped me cutting a rather long eyebrow that has started to grow in my old age! The actual match took over six months to complete, and the novel explores how the old ways of thinking gave way to the new. Unusually for a novel, it includes diagrams of the board during the course of the match, which the youthful Otaké eventually wins by five points playing black. The handicap at the time for the advantage of playing black had been set at 4.5 points. Interestingly the new style of play meant that statistical analysis of the game revealed a 6.5-point advantage for black, so the handicap has now changed. Kawabata writes of this new era, "From the way of Go, the beauty of Japan and the Orient had fled. Everything had become science and regulation."

What Kawabata would have thought of the third great revolution in the history of the game, I dread to think. The way that AlphaGo played the game to beat Lee Sedol served to break down some of the traditional conventions of how to play the game at the highest level. Indeed AlphaGo's move 37 in Game 2 has gone down in the history of artificial intelligence as the first example of creativity demonstrated by code. When the move was made, I remember journalists gasping at what a bad move they thought it was. The code had suggested placing a black stone five rows in from the edge of the board. Traditionally such a move so early in the game was regarded as very weak. During the opening moves competition was waged over the first four rows in from the edge. Your Go master would have slapped your wrists if you'd played such a move. But by the end of the game, AlphaGo had shown how to use this move to control a whole swathe of territory that ended up winning AlphaGo the game.

Creativity has been described as the ability to generate something new and surprising that has value. Move 37 of Game 2 passed all these tests. Not only that, but the move evolved out of the learning process of the code itself and wasn't preprogrammed in by a human. Indeed a human who had seen this line of code would probably have deleted it, regarding it as a weak move to make.

Some have argued that AlphaGo was not being creative. The code found this move by playing the game millions of times and coming across this occurrence enough times to amplify the connection. For a human to play the same number of games would take thousands of years. The computer certainly has an advantage as far as practice is concerned. But I don't think that lessens the legitimacy of the move's creativity. After all, our own creativity emerges from a huge amount of synaptic and neuronal activity and is probably a result of the same subconscious reinforcement. But because we can't explain the process, we feel like the conscious output is a result of some mysterious, magical thing called creativity. The bottom line is that human creativity is the result of the network of our brain just as AlphaGo's discovery was the output of a silicon network.

Since AlphaGo's triumph, the code has shown human players how to play the game in a completely new and superior way. It is now regarded as the third great revolution in the history of this ancient game. Go is an excellent game to compare to chess—but China is not without its own version of chess either.

23: Chinese Chess

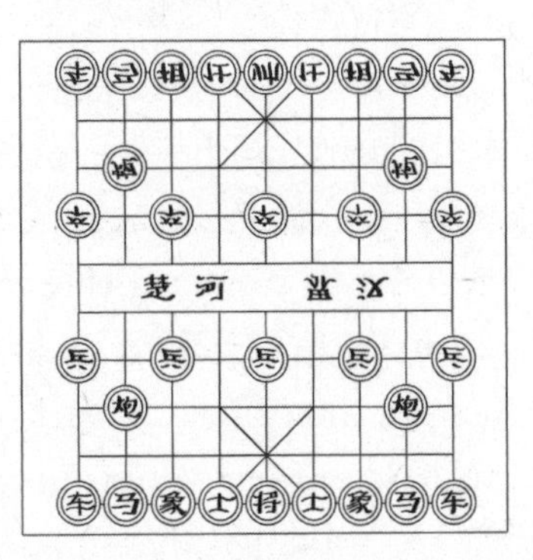

IF GO IS AN ARISTOCRATIC GAME surrounded by ritual and grandeur, one of the most popular games for the masses is Xiangqi, or

Chinese chess. In the many cafés across China, this is the game you are more likely to see ordinary people playing.

Much like the standard version of chess, it is a game of warring armies played out on a chess-like grid. But rather than occupying the center of squares, the pieces sit at the intersection points, as in Go. Another curious feature of the game is that there is a river running across the center of the board, which, depending on the board, is sometimes given a proper name. The river on the board I own is identified as the Hwang-Ho, or "Yellow River." Certain pieces in the army cannot cross the river. On each side an area marked out with diagonal lines represents the army's fortress. Confined inside the castle is the general, which is much like the king in the standard version of chess in being the piece the opposition is targeting.

The pieces are circular, with Chinese characters carved into them to denote their identities. Rather than black and white, Chinese chess has black and red pieces, similar to the way numbers in China are represented by black and red rods. The Chinese characters differ according to whether the pieces are in the red or the black army. The general shares his castle with two mandarins, who are also confined to the castle area. All these pieces can only move one square at a time. The mandarins are similar to the queen in medieval chess, before the queen was allowed to sweep across the board. The castle is guarded on either side by two elephants, flanked by two horses and then two chariots at the corners.

Elephants move diagonally across two squares but can't jump pieces and can't cross the rivers. It turns out they can occupy only seven points on the board. Horses move one step orthogonally and then one step diagonally, but unlike knights in standard chess, they can be blocked. The chariots move like castles in standard chess, riding orthogonally up and down and across the lines of the board. They are some of the most powerful pieces because of their range of movement.

The other pieces that are notably different in Chinese chess are the cannons. They move like chariots but take pieces by jumping over a piece in the space between, giving a feeling of the way that a cannon shot might take someone out on the field of battle. You can protect a piece facing a cannon by making sure the space in between is occupied. Finally, there are five soldiers assigned to each side, which are like the pawns in chess and stand at the front of the army.

The Chinese version of chess almost certainly emerged out of the game that arrived from India, perhaps some time in the eighth century, and developed its own local features independently. The first written accounts of the game date to the sixteenth and early seventeenth centuries, including a classic Ming Dynasty text dating from 1632 that documents important games. It is called "The Secret Inside the Orange" and describes the strange legend of a mysterious orange that when peeled revealed two men at its center playing chess.

Despite having bought a set when I was in Beijing, I must admit that I have never really got to terms with the Chinese characters on the pieces, so that halfway through a game I am confusing my elephants with my horses. It certainly does add an extra layer to the game play. But those who can distinguish their cannons from their chariots claim that the game boasts much of the cut and thrust enjoyed by players of its Indian cousin.

24: Pick-Up Sticks

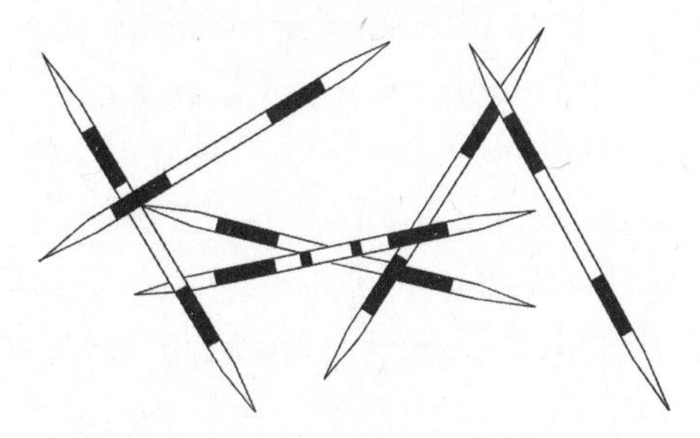

MANY ANCIENT GAMES ARE CREATED not just for entertainment but for the purpose of divination as well, as we saw in Babylon. There are some theories that the game of Go started as a method of foretelling the future according to the layout of the black and white stones as they were cast onto the board. Another much simpler children's game also appears to have had its origins in the human desire to use games to look into the future.

My sister and I often used to play pick-up sticks as kids. This simple game involved letting a load of colored sticks fall on the table and then attempting to remove individual sticks without moving any of the others, but I used to love the mix of strategy and dexterity required. Our version was called Mikado, and you'd score different points connected with the color of the stick you'd claimed. The single blue stick scored you twenty points and was referred to as the mikado, the emperor of Japan. The other sticks corresponded to mandarins and samurai, but in spite of the Japanese theme, the version I played was European in design. But that's not to say that the game doesn't have Eastern ancestry.

This popular childhood game actually has its origins in the use of sticks in the *I Ching*, the text at the center of the ancient Chinese art of divination. Just as it's thought that casting black and white stones on a table to divine the future might have been the origin of the ancient game of Go, throwing sticks on the floor played a part in ancient Chinese rituals for reading a person's destiny.

The *I Ching*, or *Book of Changes*, is an ancient Chinese text dating back perhaps as far as the tenth century BCE. At its heart is the use of hexagrams, which have different meanings as they are used in the act of divination. The symbols consist of a stack of six horizontal lines. Each line in the symbol is either solid or broken. This means there are $2^6 = 64$ different possible hexagrams. The *I Ching* contains the meanings of each of these hexagrams.

For example, the following hexagram indicates "conflict":

If the lines come out the other way around, you get

which indicates "hidden intelligence."

The connection of the *I Ching* with the game of pick-up sticks comes from the role that sticks played in determining the hexagram controlling your destiny. A bundle of fifty yarrow sticks would be used to interpret which hexagram to consult—though the finer details of how the process worked require a closer examination of the underlying mathematics.

How to Use Yarrow Sticks to Generate a Hexagram

First one stick is removed and placed down in front of you. This is the observer stick. Then you randomly partition the rest into two piles that will be laid down perpendicularly to the observer stick. A second stick is removed from the right-hand pile and placed in your left hand between the little finger and the ring finger.

Next you remove sticks in fours from the left-hand pile until you are left with one, two, three, or four sticks.

These are then placed in the left hand between the ring finger and the middle finger. Do the same for the right-hand pile. The remainder is placed in the next space along in the left hand. Now count the total number of sticks in the left hand. Note that there are forty-eight sticks after the removal of the first two sticks, which is divisible by four, so the arrangements of sticks in the left hand will give a total of five or nine sticks. This is basic modulo 4 arithmetic at work—what you are really interested in is the number of sticks remaining after division by four.

The procedure is repeated on the batches of four sticks that were removed. This will result in four or eight sticks in the left hand. Repeat again with the sticks removed on this round, again leaving you with four or eight sticks. Now score two points for each occurrence of eight or nine sticks and three points for an occurrence of four or five sticks. Add these together. You now have a number between six and nine. An even number indicates yin, a broken line in the hexagram. An odd number indicates yang, a solid line. You repeat this whole procedure for each of the six lines in the hexagram. It involves quite a meditative state to keep dividing and grouping the sticks. So it is that this act of divination rests on a fairly involved mathematical algorithmic process.

The hexagrams themselves hint at the idea of binary numbers centuries before mathematicians understood the power of six zeros and ones to count up to sixty-four. Shao Yong, a Chinese scholar of the eleventh century, made the connection, recognizing that each symbol could be attached to a number. If you write a zero for a broken line and a one for a solid line, then reading the hexagram for conflict from top to bottom would give you 111010. In decimal numbers each position corresponds to a power of ten, and the number in that position tells you how many powers of ten to take. So 234 is four units, three tens, and two hundreds. But Shao Yong wasn't working in decimal; he was working in binary.

In binary each position is simply a power of two. We saw these numbers at work in a strategy to win at the game of Nim. Working from the right of the number, 111010 would represent no units, one lot of two, no fours, one lot of eight, one lot of sixteen, one lot of thirty-two. Add these up, and the total is 2 + 8 + 16 + 32 = 58. The beauty of binary is that you only need two symbols (rather than the ten in decimal) to represent every number. If you were tempted to say I want two lots of sixteen, that simply translates into one lot of the next power of two, namely thirty-two.

The seventeenth-century mathematician Gottfried Leibniz had argued early on in his work for a means of communicating that was devoid of the ambiguities of natural language and instead totally mathematical and logical. The concept was ignored for about ten years, until he came upon a copy of the *I Ching*. He began to understand the power of binary as a means of realizing his idea to reduce all thought to mathematical language that could be implemented on a machine. He envisioned how it might resolve disagreements between philosophers, since, if they disagreed on some idea, they could simply change the problem into his formal language and then turn to the machine to sort out their differences and discover who was right. Given the way so much of our lives has moved into the digital world via our computers, smart phones, and other devices, we are perhaps closer to Leibniz's vision, inspired by the *I Ching*, than we might think.

One intriguingly unmathematical aspect of the *I Ching* is the order in which the hexagrams occur in the book. Rather than following their binary numerical interpretation, the order chosen seems to have a different rationale. The order is called the King Wen sequence after King Wen of Zhou, founder of the Zhou Dynasty. He apparently formulated this sequence while in prison. The sixty-four hexagrams are grouped into thirty-two pairs. For twenty-eight of the pairs, the second hexagram is created by turning the first upside down. The exception to this rule is for the eight symmetrical hexagrams that are the same after rotation, like the hexagram with all six lines solid. Partners for these are given by inverting each line: solid becomes broken, and broken becomes solid. This explains something of the pairing, but the order of these pairs still remained something of a mystery. There have been recent attempts to show a hidden mathematical rationale behind the sequence, including an extraordinary six-hundred-page book that seeks to relate the order to the Fibonacci sequence, implying that the sequence was known in China centuries before it was discovered by Fibonacci or the Indian musicians who found it in connection with counting rhythms.

The *I Ching* has captured the imagination of many artists over the last century. John Cage implements the *I Ching* as a method for determining how to play his piece "The Music of Changes." Philip Pullman has one of his characters in *His Dark Materials* use the *I Ching* to communicate with Dust, fictional particles associated with consciousness. Philip K. Dick employed the *I Ching* as a tool for writing *The Man in the High Castle* and then has characters in the book consult the *I Ching* as an oracle to guide their decisions.

But if you don't want to consult the oracle and prefer to leave your destiny unrevealed, then the fifty yarrow sticks could always be repurposed for a fun game of pick-up sticks.

25: Dominoes

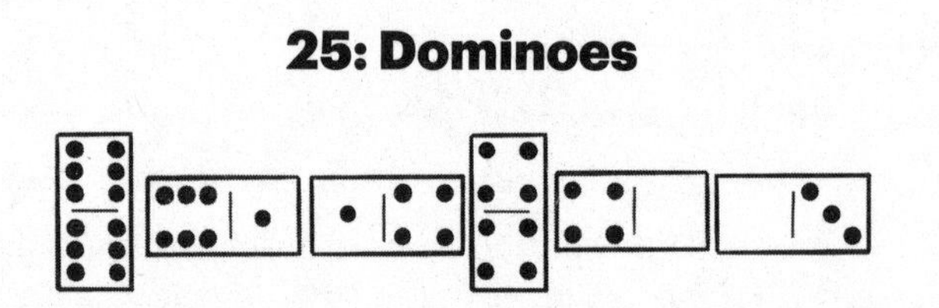

GO, CHINESE CHESS, and the *I Ching* all reflect their Chinese origins in unmistakable ways. Yet there is another ancient Chinese game with such universal appeal that you can now find it in cafés and homes across the world: dominoes.

Although there are variations in the rules, the basic premise is the same across most versions. Players are dealt a collection of tiles and in turn then try to match one end of a tile in their hand with the open end of a domino that is already on the table. If this is impossible, they take a tile from the undealt tiles. You win by eventually placing all your tiles in the train of dominoes that players are building across the table.

There is evidence of domino tiles being used in China back as far as the twelfth century. Written records tell of the game being gifted to the Song emperor Huizong in 1112 and sold by peddlers during the reign of his successor, Xiaozong, along with dice. Whoever invented dominoes was probably inspired by dice. After all each domino represents the roll of two dice (where a blank end perhaps corresponds to no throw). The blank end only appears in Chinese sets as late as the seventeenth century. The connection to dice also explains why the most popular set is marked with pips up to six. But there are sets with up to nine, twelve, fifteen, and even eighteen pips, although by the time eighteen dots are included on a single domino, it is rather tricky to discern how many pips are at the end of a tile.

It is interesting to speculate about why the pips in different sets of dominoes always come in multiples of three. One explanation is to look at the numbers in between: seven, eight, ten, eleven, thirteen, fourteen, and sixteen. Think about how to arrange pips with these numbers in a nice rectangle like we can for the pips that are multiples of three. Prime numbers like seven, eleven, thirteen, and seventeen are impossible. Even ten and fourteen only allow two rows of five or seven. The poor divisibility of ten is one reason why going decimal was such a bad decision. We should have kept to the Babylonians' base sixty, which is far more divisible. Only base sixteen admits a nice arrangement of a 4 × 4 array of pips.

You can use a domino set as a replacement for a pair of dice. By placing all of the dominoes with pips at both ends in a bag and then picking out a piece randomly, you can simulate the throw of two dice. Dominoes can also allow you to simulate throwing dice with alternative numbers of faces. Want to throw two five-sided dice? There is no Platonic solid shape that would allow you to roll a die numbered one to five. But if you just use the dominoes with pips numbering one to five in the bag, the result is the same as rolling a far more unusual set of dice.

The Math of Dominoes

How many tiles are there in each set of dominoes? Suppose you are looking to count the number of tiles in a box with pips numbered up to some number that we will call n. You would begin by laying out the tiles in rows. In the first row, put any tile with n pips. There are $n + 1$ of these, since one of the options is no pips. The next row down has all those remaining tiles with $n - 1$ pips. There are n of these because the $(n - 1, n)$ tile has already been placed in the first row. If we carry on doing this, the last row will contain a single (blank, blank) tile since all the other tiles with blanks have been placed in the rows above.

With the tiles laid out like this, we have created a triangular shape with $n + 1$ rows—but how many tiles are there in total? To find out, we can play a clever trick. Take a second set and make a similar triangle of dominoes, except this time build the rows from the bottom to the top. Now push these two triangular shapes together to make a rectangle of dominoes with $n + 1$ rows but with $n + 2$ dominoes in each row. So the total number of dominoes in this rectangle is $(n + 1) \times (n + 2)$. But this was two sets of dominoes, so the number of dominoes in a single set is $(n + 1) \times (n + 2)/2$.

Taking our standard set with six pips and setting $n = 6$, the formula tells us that there are $\frac{7\times 8}{2} = 28$ tiles in a standard set. If you are

playing with four players, this is usually too small to play a satisfying game, which is why the nine-pip and twelve-pip sets will often be used in multiplayer games.

And now, no matter how large those sets get, you have a method for counting them. This is useful if you think a domino tile might have gone missing down the side of the couch.

Chinese dominoes differ slightly from our modern versions. There are two different suits: the civil suit and the military suit. The civil suit consists of eleven different patterns: 6-6, 1-1, 4-4, 1-3, 5-5, 3-3, 2-2, 5-6, 4-6, 1-6, 1-5. The military suit consists of the remaining patterns, a total of ten.

The civil suit is doubled to make a total of thirty-two tiles in a Chinese box. The tiles in the civil suit each have a Chinese name, and these names indicate a rank order to the tiles. The 6-6 is *tin* (天, heaven), 1-1 is *dei* (地, earth), 4-4 is *yan* (人, man), 1-3 is *ngo* (鵝, goose, or 和, harmony), 5-5 is *mui* (梅, plum flower), 3-3 is *cheung* (長, long), 2-2 is *ban* (板, board), 5-6 is *fu* (斧, hatchet), 4-6 is *ping* (屏, partition), 1-6 is *tsat* (七, long-leg seven), and 1-5 is *luk* (六, big-head six).

The idea of dominoes having suits and ranks within these suits, some have suggested, might have been the spark for card suits. The domino games that I grew up playing are all about putting pieces down in turn, trying to match tiles and extend the train of dominoes across the table. Sometimes you haven't got a tile that matches either end, and you have to take a penalty tile from the pile of undealt tiles. But here's a question for you: Is it possible in all the sets available to make a train using all the tiles?

This turns out to be related to one of the great puzzles of mathematics: the Bridges of Königsberg. The seventeenth-century residents of the Prussian town of Königsberg used to entertain themselves by playing a game. The town was divided into islands by the river Pregel. There were seven bridges that connected different bits of the town. The game was to find a way around the town so that you crossed each bridge once and once only.

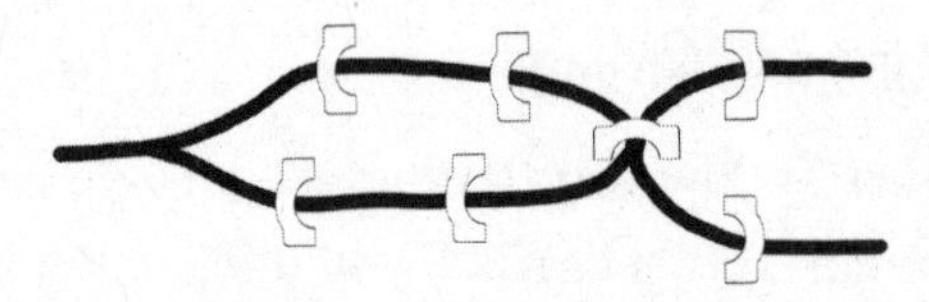

However hard they tried, the residents were always left with one bridge they couldn't cross. Not until Swiss mathematician Leonard Euler brought his mathematical perspective to bear was the puzzle finally resolved. Instead of the map of Königsberg with all its intricate detail, he analyzed an alternative map that picked out the essential qualities of the problem. The problem of the Bridges of Königsberg could then be identified with the problem of drawing the new map without taking your pen off the paper and without running over a line twice.

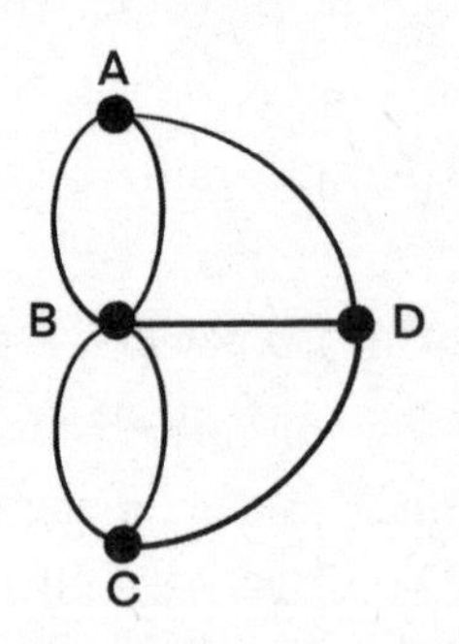

The graph crystallizes the essential information that is needed to solve the problem. Euler then observed that in any network, except for your initial and final location, every node in the graph must have an even number of edges emanating from it if a path around the graph is possible. If it is possible to find a path round Königsberg to solve the residents' game, there must be at most two nodes with an odd number of edges. The graph of the Bridges of Königsberg has four nodes with an odd number of edges, so it is impossible to navigate the town crossing each bridge once and once only, just as the residents discovered by trial and error.

Euler proved that a path is possible provided there are two or zero places with an odd number of lines emerging from the points. If there are two, these represent the starting and finishing points of the journey. Alternately, if there are none, then your journey will return you to the point where you started.

So how does this help with our domino problem? The islands in the city map are replaced now by the different numbers that can appear on the end of the domino pieces, and the bridges are the domino pieces connecting these numbers. We can construct a chain consisting of all the domino pieces provided we can draw a picture around this map crossing every domino piece once and once only.

Let's start with the six-pip domino set. The key is to count how many bridges or dominoes emerge from every point in the graph, so we need to count how many dominoes have each number at one end. Take the six pips. This can be paired with blank up to six. So that looks like seven bridges. But notice that the double six represents a loop, so it actually contributes two lines coming into the point. So each point has eight lines coming out of it.

By Euler's theorem, this means that there is a path around the dominoes because every point has an even number of lines emerging from it. This will be true for any set of dominoes where the largest number of pips is even. But take the set with nine pips, and everything changes. Now there is an odd number of lines emerging from every point. Euler's theorem asserts that it is impossible to make a chain out of all the dominoes in a nine-pip set.

Although these tiles began their journey centuries ago in China, it is in Latin America and the Caribbean that the game is most widely played today. A game of dominoes was even responsible for the name chosen by a town in Florida. In 1867, Henry T. Titus arrived in a community in Florida known as Sand Point. He was keen to develop the land into a town and began to build a hotel, churches, and a courthouse.

But he had competition for ownership of the town. Captain Clark Rice also had his eye on it. Both wanted to name the new town to stamp their claim to the community. They decided to use a game of dominoes to determine who could claim the right to name the town. Titus won, and to this day the town in Florida is known as Titusville. Had the tiles landed differently, it could have been Riceville.

Universally recognized across the world as dominoes may be, another set of tiles created for gaming purposes has become unmistakable in its connection to Chinese culture.

26: Mah-jong

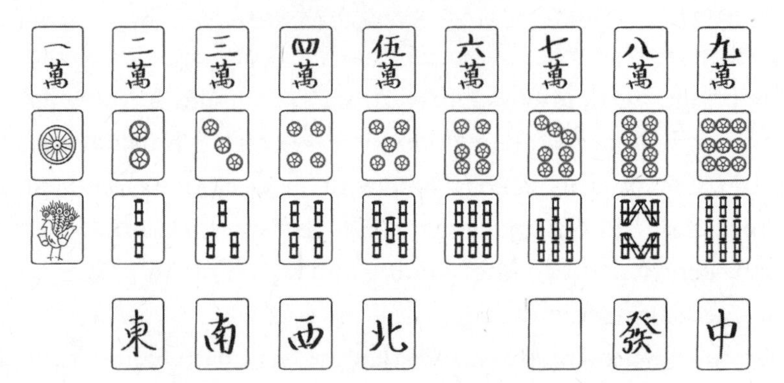

ONE OF MY FAVORITE CARD GAMES as a kid was rummy. In this game you collect either three of a kind or runs of cards in a suit. Although it looks very different, a similar process is at the heart of one of the most iconic of Chinese games: mah-jong. Instead of cards, players collect ornately decorated tiles traditionally made from bone and backed with bamboo.

The playing of mah-jong has such a distinctive soundtrack that it always takes me back to China when I hear it. It is the noise of the mah-jong tiles clacking around the tabletop as the four players mix them up in the game's version of shuffling the pack. In fact, one of the names that the game goes by, maque, is thought to originate from the word for sparrow, because the tiles clicking against each other sound like a flock of twittering sparrows.

Mah-jong is not an ancient game and only appears for the first time around Shanghai in the mid-1800s. Nevertheless it has won its place as a central part of modern Chinese culture. Mao Zedong believed it to be one of three great contributions that China has made to the world: "Mah-jong is a philosophy. Through it, one understands the relationship between chance and inevitability; it's also dialectical... [E]ven if you have the worst hand, as long as you are strategic and methodical, the inferior will become superior; weakness will become strength."

The game was so popular that some Communist Party leaders began to worry that mah-jong addiction might distract the country from progress. Despite Mao's praise for the game, to try to avoid the corruption of the country through gambling on the tiles, the People's Republic of China banned the playing of the game for many years. It was eventually legalized in 1985, but only for playing, not gambling.

In an essay in 1930, Chinese scholar Hu Shi wrote, "The increasing complexity of the game made it even more addictive, resulting in Chinese people, whether male or female, rich or poor, spending all their time on the 136 tiles, all year round." He estimated that if there were "a million mah-jong tables in China, then even if everyone only plays eight rounds, that adds up to four million hours, which is 167,000 days wasted. Not to mention the money won or lost, and the waste of energy." The game became popular in the West too during this period. Indeed, in 1923 mah-jong sets outsold radios in Britain, an impressive feat given that the radio, or "wireless," was the principal source of entertainment for most houses at the time. But once you start playing the game, you can quickly begin see why it is so addictive.

The aesthetic of mah-jong is irresistible, starting with the sensual clacking of the tiles on the table as they are shuffled. The game begins with 136 tiles stacked together to make four walls, two tiles high atop the mah-jong mat, reminiscent of an ancient Chinese fortress with the courtyard in the center. There is something satisfying in the way that each mah-jong tile sits in your hand, like a perfect pebble picked from the beach, and each tile sports a beautiful array of symbols.

The game appears to tell a story of magic and ancient times. The tiles include three dragons: red, green, and white. There are also the four winds blowing from north, east, south, and west. Some sets include eight bonus tiles depicting the four seasons and four flowers that conjure up the gardens of the east: plum blossom, orchid, chrysanthemum, and bamboo. The bread and butter of the tiles consists of the tiles of the three suits: circles, bamboo, and characters numbered from one to nine.

The word *mah-jong* literally means "sparrows." Some theorize that the name of the game has to do with the tiles that were given to reward those who killed sparrows to stop them eating grain. The number of bamboo sticks on your tile reflected the number of sparrows you had killed and could be exchanged for rewards commensurate with the number of sticks. This perhaps explains why the "one of bamboo" is in fact a picture of a bird. The circles could refer to the holes left behind by gun shots. Even the dragons have a sparrow interpretation: the symbol for the red dragon actually means "shot sparrow."

I find the character suits particularly enjoyable because they offer a fun challenge to identify the Chinese characters for each number. Some sets have Western numerals to help you play, but my set doesn't pander to non-Chinese players. Below each of these characters on the tiles is an additional Chinese symbol, which is the character for a *myriad*, or ten thousand. So actually the tiles are supposed to represent a much greater number—an indication that they were originally used as some sort of gambling token.

The ancient Chinese used a system of sticks to do calculations and represent numbers. There is a curious twist here: the numbers are represented sometimes by horizontal lines, other times by vertical ones. This helps in distinguishing when rods were being used for units, tens, or hundreds. For example, three vertical sticks could represent 3, but they might also represent 21 or 12 or 111. How could you tell if two of the sticks were meant to be in new columns? In order to keep track of sticks moving from units to tens, the sticks would be rotated by ninety degrees. So 111 would be represented by one vertical rod followed by a horizontal rod to indicate a shift to tens and then another vertical rod to indicate a shift to hundreds.

𝍩	𝍪	𝍫	𝍬	𝍭	𝍮	𝍯	𝍰	𝍱
1	2	3	4	5	6	7	8	9
𝍠	𝍡	𝍢	𝍣	𝍤	𝍥	𝍦	𝍧	𝍨
1	2	3	4	5	6	7	8	9

The Chinese were also interested in negative numbers, especially from a financial point of view. If you owed someone money, then that debt needed to be represented by a number. This led to the invention of negative numbers. This made it easier if, for instance, I owed three coins to my neighbor, but then I earned four coins. The introduction of negative numbers allowed me to easily calculate that my net worth was one coin: $-3 + 4 = 1$. The sticks that the Chinese used to keep track of numbers would be colored black and red to denote the difference between money owed and money owned. Hence we have the saying "going into the red" to denote the fact that you owe money. Curiously, in China the black sticks denoted negative numbers. Somewhere along the Silk Route the colors got swapped.

Mah-jong is designed for four players who sit at the four sides of the square mah-jong table. These places correspond to the four compass points, which play an important role in the game. At its heart it is a game of collecting akin to the card game rummy that I liked as a kid. There are four copies of each tile. This means that one can try to collect tiles of one particular symbol. Collect three identical tiles, and you've got a *pong*. Get all four, and it becomes a *kong*. You can also aim for a sequence of three tiles of the same suit, called a *chow*. For example the three, four, and five of bamboo form a *chow*.

You start with thirteen tiles in your hand, and during the game you take new tiles from the wall or your opponents' discarded tiles and then discard your own unwanted tiles with the aim of having a hand made up of pongs, kongs, and chows. Twelve tiles, for example, allows you to have a hand of four pongs. But what about that leftover thirteenth tile? You go mah-jong by picking up a tile that makes an identical pair with this odd tile, which is called an *eye*. Your final hand consists of fourteen tiles since you don't discard when you go mah-jong. If you succeed in getting all four tiles of one sort, a kong, then the number of tiles can be more than fourteen. The maximum number of tiles is eighteen, corresponding to four kongs and a pair.

These are the basics of mah-jong, though one of the other features that I love about this game has to do with the other, more exotic hands you can end up with, which often come along with beautiful names. These include the Thirteen Orphans, the Heavenly Twins, the Thirteen Unique Wonders, and the Dragon's Breath, among others.

To assemble the Thirteen Orphans, instead of collecting things in groups, everything sits on its own. You need one of each wind, one of each dragon, and then the one and nine of each suit, plus one extra of any of these. You have to be pretty brave to assemble these pieces because it's really running in the opposite direction of a classical mah-jong hand. Still, these rarer hands score you more points in the game. It is the accumulation of points that you score as you play more rounds that ultimately wins you the game.

When I was in China, I couldn't resist diving into the market in Guangzhou and picking up a box of tiles together with a mat on which to play the game. It is not a fancy set—just 136 tiles without any bonus tiles. The tiles are acrylic, white with green tops, and come in a plastic blue box covered in Chinese characters I don't understand. It was a heavy piece of hand luggage to take on the plane back home, but it was worth it, since my family always enjoys getting out the tiles and building the walls. My girls were also pleased that their mah-jong training meant they actually understood an important scene in one of their favorite movies, *Crazy Rich Asians*. In the crucial mah-jong scene, a rich mother sits in the dominant East position representing Asia, while her son's not-so-rich girlfriend, who was brought up by a single mother who fled China to the United States, sits in the West seat representing America.

As the game proceeds the girlfriend understands the tiles that the mother is collecting, so when she picks up the eight of bamboo, she knows it is the last tile that the mother needs to go mah-jong. But we see that this tile will also allow her to win the game. It is at this point, before she's decided whether to keep the tile or not, that she tells the mother that her son has proposed to her. But before the mother can react, the girlfriend tells her she turned the son down.

Dumbfounded, the mother declares, "Only a fool folds a winning hand." The trouble is that in this game of love, there is no winning move. The son loses his family if he marries or loses the love of his life if he follows his mother's advice. The girlfriend saves him from making either losing move by rejecting his proposal.

The girlfriend then discards the eight of bamboo, which allows the mother to triumphantly go mah-jong. But when she sees the girlfriend's hand as it is laid down, she realizes that the eight of bamboo would have allowed the girlfriend to beat her and understands the true sacrifice that she has made. Given that we know from the opening that the girlfriend is actually an economics professor specializing in game theory, she probably knows when folding a winning hand is the optimal choice in the bigger game she is playing: to win the son. It is a brilliant scene but remains mysterious unless you know your mah-jong.

Other games evolved alongside mah-jong in China that may not have become as culturally important but stand out for their uniqueness. Although mah-jong itself is not an ancient game, it might have evolved from a card game that could be the origin of all card games across the planet.

27: Zi Pai, Khanhoo, and the Origins of Playing Cards

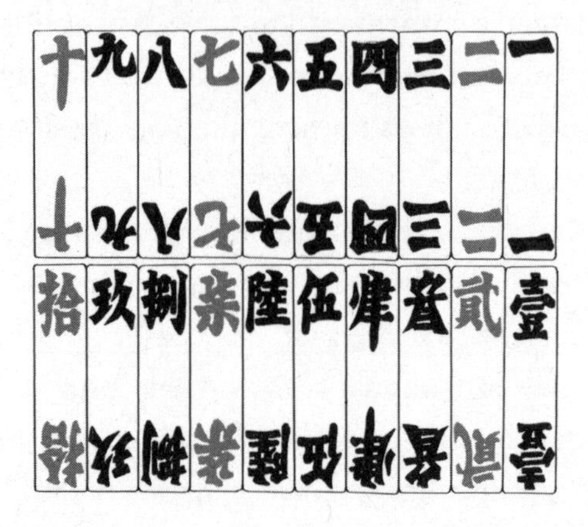

ONE OF MY FAVORITE GAMES from all of my travels is a strange set of playing cards that I've never seen anywhere outside China. I first came across them after traveling with my family to Yangshuo, a small city surrounded by dramatic karst mountains, winding rivers, and beautiful scenery. We spent the day cycling by the famous Li River and watching the fisherman using cormorants to catch fish. But only after stopping at the local café for lunch did I spot the thing that trumped the other sites we'd seen during the day. A group of men were furiously playing a game with a set of cards that I'd never seen before: beautiful oblong cards covered in Chinese characters. I sat for ages watching them play and trying to decode the rules of the game. Eventually a young woman took pity on me and showed me how the game worked. I still have the now rather crumpled paper on which she kindly wrote out the rules for me.

The game these men were playing turned out to be a sort of early forerunner of mah-jong. The cards were called Zi Pai, which translates as "character cards," and were so beautiful that I felt convinced that people back home would jump at the opportunity to buy the cards and learn the game they were playing. I came home with a bag full of packs of Zi Pai. My entrepreneurial vision didn't come to fruition, but I still love playing with the cards that I bought that day.

The deck consists of two suits numbered from one to ten. The first suit uses the ordinary Chinese characters for numbers, like those on the tiles in mah-jong. The second suit has the numbers written in formal Chinese characters, and these are most challenging to keep track of. There are four of each card, making a total of eighty cards in each pack. (My pack has an eighty-first card with a strange Chinese character on it whose meaning I don't know.) The number two, seven, and ten cards are printed in red, while the others are printed in black. These red cards give a hint as to how to play a game often played with this pack and sometimes known as 2-7-10.

The game is played by three players who each get twenty cards. The aim of the game is the same as in rummy or mah-jong: to get three or four of a kind. The exception is that getting 1-2-3 or 2-7-10 of a suit is another way of melding the cards and scores you more points.

These cards are probably descendants of what many regard as the oldest playing cards in the history of games, known as *money-suited cards*. It seems that the cards were originally not actually part of a game but rather used to keep track of bets on a game. The cards consisted of three suits (sometimes with a fourth suit added). The first suit depicted circles, symbolizing coins, which in China have a hole in the middle. The second suit depicted strings. These were pictures of one hundred coins threaded together on a string through the hole in the middle. The third suit denotes ten thousand coins and simply shows the number of sets of ten thousand the card represents by depicting the Chinese character corresponding to that number. The designers of the cards were essentially representing the discovery of the place number system in card form.

One of the earliest records of cards being used for games comes in a report of a legal case brought against two gamblers in July 1294. "We caught Yan Sengzhu and Zheng Zhugou playing cards, and have also found wood blocks to print cards. Each person has admitted to the truth of the accusation." The report goes on: "Dispatched to the Ever-abundant Treasury for deposit the nine cards (zhipai) that were about to be destroyed." It is possible that the money-suited cards were originally used like chips at the casino to denote gambling bets.

At some point it seems likely that these cards started to be used in a game in their own right. It is interesting that while trick-taking games are at the heart of card games in the West, collecting cards of the same type in the style of rummy seems to be a trait corresponding to many games in the East.

We see a very clear connection between these money-suited cards and the main suits used in mah-jong. Coins become circles. Strings become bamboo shoots. And the Chinese numbers counting myriads become the character suit.

An intermediate game called Khanhoo emerged in the Ming Dynasty. This uses a version of these money-suited cards but adds three extra wild cards: red flower, white flower, and old thousand. It is possible that these wild cards could have been the forerunners of the dragons that appear in mah-jong. Very intriguingly there is a Mexican game called Conquian, one of the earliest-known rummy games, which sounds sufficiently like Khanhoo that one might wonder at how all these games are connected. Games can act like a historical fingerprint revealing how different cultures interreacted in the past. As we bid farewell to China's impressive collection of games, we head back to sea as we resume our journey, this time to Japan.

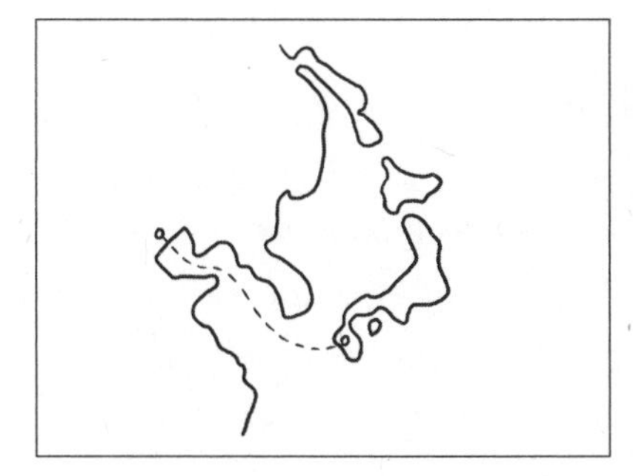

CHAPTER 6

THE EAST CHINA SEA

Cards

AS WE RETURN TO SEA ONCE AGAIN, we have now encountered card games of such a variety of different types that the time is right to explore the larger world of cards and their impact on humanity. In the Ganjifa cards we saw in India and the money cards we discovered before leaving China, I am fascinated to see two different cultures concoct games that reflect such distinct and different characteristics. Ganjifa cards were used to play games where players win tricks, taking the cards of their opponents. The resulting games are aggressive and immediate. China's money cards are designed for players to focus on collecting and building with their cards. It takes time and patience to build up the hand that wins the game. Much like the differences between chess and Go, we see diverging approaches that reveal much about the differing cultures and about humanity more generally.

Most fundamentally, I regard a pack of cards as one of humanity's most extraordinary inventions. A small cardboard box compact enough to fit in your pocket contains such a variety of different games that one could spend a lifetime learning them all. Of all the games in this book, it is a pack of cards that I would bring to any desert island I might find myself stranded on in my journey around the world.

28: Whist

ALTHOUGH MY ROUTE AROUND THE world wends its own course, independent of Phileas Fogg's, the one thing his journey has in common with ours is that he too whiles away the long train and boat journeys by playing games. One game in particular that he was obsessed with, as were many of his generation, was whist.

It is during a game of whist at the Reform Club that Fogg takes up the challenge of circumnavigating the globe in eighty days. Frankly he isn't interested very much in sightseeing or experiencing the different countries he visits; rather he is focused simply on achieving the task of the bet he has made. While amazing scenes pass by the window of the train, Fogg is much happier to have his head in a hand of cards.

As Jules Verne wrote, "Mr. Fogg played, not to win, but for the sake of playing. The game was in his eyes a contest, a struggle with a difficulty, yet a motionless, unwearying struggle, congenial to his tastes."

As the city of Bombay comes into view, Fogg is busy finishing a mammoth round of whist consisting of thirty-three rubbers. In the final game, he and his partner achieve the rare feat of winning all thirteen tricks available—a far more exciting sight for Fogg than being on deck watching the grand arrival into the Bay of Bombay.

The game of whist was popular for centuries before Fogg's obsession with the game. Indeed, the sandwich owes its origins to the Earl of Sandwich's own obsession with the game. He was so immersed in a protracted series of games of whist played at the tables of a club in Bath, a major center of whist playing in the eighteenth century, that he was too engrossed to take a break for food. Instead he demanded that slices of roast beef be brought to him at the table, placed between two slices of bread to prevent his hands and cards from getting greasy. Others saw the appeal: "I'll have what Sandwich is having!"

The game of whist is a trick-taking card game for four players. All fifty-two cards are dealt out with each player receiving thirteen cards. The game has its origins in the game of Ganjifa, but while the Indian forebear dictated rather uncreatively what order cards should be played in, whist allows players to choose to play high or low cards as they wish. The game also has the important feature of playing in teams of two. You're playing not just the thirteen cards you can see in your own hand but also the thirteen unseen cards in your partner's hand across the table.

The other new feature is the role played by trump cards. As the fifty-two cards are dealt, the last card is turned over and held by the dealer, its suit becoming the trump suit for that round. This means that this suit assumes an extra power, such that even if you aren't able to follow the suit of the card led during a trick, you can use a card from the trump suit to win the trick.

The idea of trump cards actually comes from games played in the fifteenth century with tarot cards (which I shall come to later), where a fifth suit, the Major Arcana, assumes this powerful role. In a conventional pack with only four suits, players began to have the idea of nominating one of the suits to assume this extra power in the game.

In playing whist, teams would score points for the number of tricks over six made during a round. The winning team originally was the one that got to ten points. But if we return to the card tables of Bath at the beginning of the nineteenth century, we find Lord Peterborough losing a huge amount of money at the tables. His friends agreed to offer him a fast way of recouping his money, cutting the winning number of points down from ten to five. The players so enjoyed this racier version of the game that five-point whist, or short whist, became the dominant way of playing from about the mid-nineteenth century.

The exciting new element of game play in whist is that players realized that they could use the cards played during a trick to communicate information about the other cards in their hand. The realization led to a new science about how to play the game in the most effective manner. As James Clay, one of the important nineteenth-century analysts of the game, wrote, "Whist is a language and every card played is an intelligible sentence." Cards became code.

This emerging scientific approach to the game coincided with the Enlightenment, the philosophical movement placing reason at the heart of the pursuit of knowledge. Whist players believed that one didn't need to leave the game to chance and could bring reason to bear on how best to play the game. A group that would meet at the Crown Coffee House in Bedford Row in London around 1728 began to develop the science of the game. The coffeehouses were the forerunners of the gentlemen's clubs, like the Reform Club, where Fogg played his whist. The science of whist proposed paying particular attention to how to play your cards to communicate to your partner what else you had left in your hand and what card you might want them to play next.

For example, a convention called the Blue Peter, named after the flag used to signal messages in the navy, allowed players to ask their partner to lead with a trump. You did this by playing an unnecessarily high card in the current trick being played. Your partner would realize that it was a bit strange to waste such a high card when a lower-value card would do and understand that you were communicating the wish for trumps to be led with next.

These conventions became increasingly convoluted. In time, pages and pages were written and tutors were sought in order to prepare players to play the game at the standard that was demanded. For example, Foster's "Rule of Eleven" requires a player to deduct from eleven the value of a card played by their partner in order to ascertain how many higher cards of that suit were lacking from their hand.

The only trouble with all of this was that you had to waste time playing tricks before much information could be communicated. Players enjoyed this side of using cards to gain insights into the missing cards. But they wanted some way of doing this without having to play the game. To satisfy this desire, whist evolved into what would become one of the most popular card games of all time.

29: Bridge

BRIDGE ADDS A NEW TWIST TO WHIST: bidding takes place before any cards are played. After their hands are dealt, players take turns to bid on how many tricks they think they will make above six tricks if a particular suit is trumps. For example, a bid of two spades indicates a belief that your team can make eight tricks if spades are trumps. The bidding round is one of the really fun parts of the game of bridge, which happens before a card has even been played. In order to outbid someone who has already declared, you have to claim that you think you can make more tricks than the current bid. There is a hierarchy of suits: spades beat hearts beat diamonds beat clubs, such that bidding two hearts beats two clubs.

The bidding has become a place where you might actually communicate information about the cards in your hand rather than your actual goal. A whole host of different conventions have evolved: the Blackwood, the Stayman, the Jacobi Transfer, and many more. The Blackwood is used to work out how many aces and kings you collectively have, which could allow you to make a bid to win all the tricks if you have them all. Bidding five clubs, five diamonds, five hearts, or five spades is just a signal about whether you have zero, one, two, or three aces in your hand. You have to be careful using these conventions. You don't want to end up having to play a hand for a bid simply meant to communicate information to your partner rather than to be a statement about the tricks you think you can realistically make.

The bidding means that players can learn things about their opponents' cards that in the game of whist would have required seven or eight tricks to gain the same knowledge. Some believe that even the name *whist* has its origins in an obsolete word meaning "hush." Whist required such attention to the cards that it was often played in silence. One of the reasons that bridge seems to have caught on is that people enjoyed the element of conversation that is involved in the bidding stage. This aspect of the game, it is thought, attracted more women to the card tables.

This extra component of the game, added at the beginning of the twentieth century, has seen bridge usurp whist as the game of choice at the card tables of the world. At school it was certainly the game of choice among my friends. All my break times at school were taken up with playing bridge, provided we could find a secret hideaway where we wouldn't be beaten up as nerds. Generally the storeroom cupboard in the music department provided a safe house for our bridge binges.

In November 2011 a news story broke in the United Kingdom about what seemed like the most unlikely hand of cards dealt in the history of the game. A group of pensioners in Warwickshire met regularly to play, but after Ron Coles had dealt the cards for their first game, he was rather shocked to pick up his hand to discover all thirteen spades. But he could see by the reactions on his friends' faces that they were also rather stunned by their hands. It turned out that one had all the hearts, another all the diamonds, and the fourth all the clubs—an incredible coincidence whose underlying mathematics are worth a closer look.

The Math of the Perfect Deal

There are fifty-two cards in the pack, so that means there are $52 \times 51 \times 50 \times \ldots \times 3 \times 2 \times 1 = 52!$ different possible arrangements of the cards. How many of these arrangements would result in such a deal of four hands where all the cards of one suit end up in the same hand? To work this out, we start by laying the first card on the table. This can be any of the fifty-two cards in the pack. It's just fixing the suit of the first hand. The next card, which begins the second pile, must be a card of a different suit from the first card. So that is a choice of $52 - 13 = 39$ cards. The third card needs to avoid the first two suits, so this is a choice of twenty-six cards. And finally we have a choice of thirteen cards for the last pile coming from the remaining suit.

Now we have determined the suits of the four piles; from here on the cards must be arranged according to the suits in the piles. There are twelve cards left of each suit, and they can be arranged in any order, which means that the number of ways of arranging the cards to achieve this perfect hand is

$(52 \times 12!) \times (39 \times 12!) \times (26 \times 12!) \times (13 \times 12!)$.

So the probability of the perfect deal occurring out of all the possible deals is this number divided by 52! That translates into the chances of the Warwickshire pensioners dealing such a hand being 1 in 2,235,197,406,895,366,368,301,559,999. To get a sense of the size of this number: if everyone on earth had dealt cards at one deal per second for the lifetime of the current universe, we still wouldn't have dealt that many hands.

This would seem a profoundly rare occurrence in bridge history—yet, bizarrely, it was not so unique after all. When my fellow mathematician Peter Rowlett heard this story in 2011, he remembered a similar news story breaking a few decades earlier. Sure enough, looking back he found that the same thing had happened to another group of pensioners in Bucklesham in Suffolk in January 1998.

Digging further back through the newspapers, he uncovered the same miracle hand had been dealt in January 1978 in Milwaukee, in April 1963 in Wyoming and a week before that in Illinois, in March 1954 in Rhode Island, in July 1949 in Virginia, in March 1935 in Pennsylvania, in November 1929 and July 1928 in London, in 1927 in Maine and in Bedford in England, and in 1892 in Brighton; the first ever recorded example was in Calcutta in 1888.

For an event that is meant to be so rare, it's happened a lot of times! What's going on? The point is that card players tend not to shuffle the cards more than a couple of times, thinking that a couple of riffle shuffles should be enough to make the pack random. A riffle shuffle is when you split the pack in two and then, holding a pile in each hand, interleave the two piles by letting the cards fall from between fingers and thumb.

Most of the time this results in the cards not being perfectly interleaved but a few cards at a time landing from each pile. But there is a version of this shuffle where the cards are perfectly interweaved, one card at a time from each pile. This is called the perfect shuffle. If you start with a new pack of cards, where the suits are arranged in order, and then perform two perfect shuffles, the effect is to rearrange the cards such that every fourth card is from the same suit. Deal this arrangement out in a game of whist or bridge, and bingo, you've got four hands, each with cards of the same suit.

So the perfect bridge or whist hand is not as rare as one might first expect. Given that many clubs will play with new cards arranged in suits and a perfect shuffle is easier to do with a clean set of cards, perhaps it's more surprising that the perfect bridge hand only makes the news every couple of decades.

How many times should players shuffle the pack in order to ensure that the cards are randomized as much as possible? If you are doing a riffle shuffle and there are l cards in your left hand and r cards in your right, then a sensible model is to say that there is an $\frac{l}{(l+r)}$ probability that the next card is going to fall from your left hand. An analysis of the mathematics of this shuffle reveals that you need to shuffle the pack seven times for it to become random. Any less than this, and the pack retains information from the previous game.

But be careful who is shuffling the pack. Because magicians actually learn how to do a perfect shuffle and can exploit the fact that if you repeat a perfect shuffle eight times (no mean feat), the pack magically (or mathematically) returns to its original order. The perfect shuffle is a bit like rotating an eight-sided coin. Each shuffle is like moving the coin around an eighth of a turn. After eight shuffles, just as the coin has returned to its original position, the deck is just as it was before you started shuffling. For the punter looking on, eight shuffles would appear to have completely randomized the pack. But the magician knows exactly how the pack is arranged and can now exploit that fact to wow the audience.

In the casinos of Las Vegas, machines are often used to shuffle the cards in order to avoid any cheating by human hands. But believe it or not, machines too can fall short. A colleague of mine, Persi Diaconis, a professor at Stanford specializing in randomness, is often asked by casinos to come in and assess their new machines. I remember him telling me about this wonderful new contraption that a local card-shuffling company in Vegas was very proud to have built. The machine had ten shelves. A deck of cards goes into the back, and then the cards are allocated to one of the ten shelves using a random-number generator. The machine runs up and down the contraption, quickly dealing the cards into the ten shelves. Further, when a card is put on a shelf, it is either put above or below the current pile at random. And then the piles from the ten shelves, which each have about five cards on them, are stacked in a random order. The machine does this once, and that's the shuffle. It all looked very slick. The company had built twenty-five machines, ready for distribution in the casinos, after investing several million dollars. They were pretty confident that the machine was randomizing the pack. But to be sure, they asked Diaconis to take a look.

The trouble was that Diaconis's math told him that the machine was leaving far too much order in the pack. For the CEO of the company to understand how much information was being left in the pack, Diaconis came up with the following explanation. If the pack is random and I ask you to guess the next card as I turn the cards over in turn, how many cards would you expect to get right? Well, there's a $\frac{1}{52}$ chance you get the first card right, then $\frac{1}{51}$ you get the next one right, given you've already seen one of the cards. Running through the pack, this results in your being able to correctly guess

$$\frac{1}{52}+\frac{1}{51}+\ldots+\frac{1}{3}+\frac{1}{2}+1=4.538$$

cards. This sum of fractions is part of something called the *harmonic series*. Intriguingly, if you add all the fractions like this, the answer becomes infinite. In a conventional deck of cards, you should actually be able to correctly guess between four and five cards out of the fifty-two.

The trouble is that if the pack was shuffled using this beautiful new contraption, his analysis of the mathematics of the machine revealed that the number of cards you could guess would increase by another five cards. The company realized that despite the engineering being a thing of beauty, the machine would still not be suitable for use in the casino.

Diaconis's other obsession is with games of patience, or solitaire, as they are known in the United States, and the question of how often one can be expected to win. Perhaps the most-played game of patience is Klondike or Idiot's Delight in which seven piles of cards are dealt with one card in the first pile, two in the second, up to seven in the last. The top card of each pile is turned over. The remaining cards are sorted through so you only see every third card. Acceptable moves are to place one card on another if the card you are moving is of a different color and one less than the card it is being put on top of. So a red seven can go on a black eight, and a black jack can go on a red queen. The aim is to pull out the aces, followed by pulling out any twos and placing them on available aces, and continuing through threes, fours, and so on, until you have cleared all the cards.

In Vegas you can buy a deck of cards for $52 and instead of continually cycling through the remaining pack seeing every third card, you are allowed to see every card, but only once. For every card that you put up, the casino pays you $5.

Despite this game being played since its first introduction around 1780 and originally being packaged with every copy of Microsoft Windows, no one knows the average success rate for clearing the deck. Considering you can make $5 a card in Vegas, it's worth knowing whether the odds might in fact be in your favor to beat the house. There are enough complications involved in the rules that even Diaconis is foxed by how to calculate the average success rate. He has been collecting data over the years, and it looks like the average might be that you win 15 percent of the time. But proof still remains elusive.

30: Spades, Hearts, Diamonds, and Clubs

THE GANJIFA CARDS THAT I ENCOUNTERED in India (or rather at the Victoria and Albert Museum) could have been the fusion of two packs of cards being used in the Mamluk sultanate in Egypt. But they did not remain in India and northern Africa; they also made their way west to Venice, where the suits were translated into coins, cups, swords, and polo sticks. These originated as symbols used in the Mamluk courts to denote four important offices: the exchequer, the cup bearer, the commander of the palace guard, and finally the polo master. They may help us explain the association of modern suits with different strata of society: clubs with the peasant class, spades with the military or nobility, diamonds with the merchant class, and hearts with the clergy.

The origins of all these symbols seem to go back to the Chinese money cards made up of three suits—coins, strings of coins, and myriads of coins—which, as I've explained, could be the origins of all playing cards. The first level of these Chinese money cards accounts for the coin suit we see in nearly all early packs of cards. It might be that the string of coins became the polo stick, the idea being that you could thread the coins onto the sticks. One theory for the cups is that the Chinese symbol for a myriad looks like a cup when turned upside down. The swords come from the tens of myriads cards because the symbol for ten in China is a cross that looks rather like a sword.

It seems that Europeans didn't really have much idea about polo, and so the sticks morphed into a club used to strike someone. We now have the suits in place for Italian and Spanish packs of cards: coins, swords, clubs, and cups. These are the cards that I've encountered while hiking in the Dolomites, where card games like Scopa are the evening pursuit after a long day in the mountains. But once you cross the Alps into Switzerland, the suits change. Cups, coins, clubs, and swords mutate into roses, bells, acorns, and shields. These suits appear in Europe as early as 1450. As the cards move into Germany, the roses become the hearts we now recognize, while the shields become leaves. When we get to France, the suits finally correspond to the modern pack most recognized around the world. Here the coin and bell of previous packs have become the diamond. The leaf becomes the spade. The acorn mutates into a clover symbol, but we seem to have retained the original name, club, that corresponded to the first mutation of the polo stick from the East.

The changes that occur in the suits from one pack to the next probably don't have a clear rationale. The different symbols are likely just as much a result of the whim of the rich family commissioning the cards. For example, the bell symbol is thought to be related to the bells worn by hawks in falconry, an activity that is the preserve of the rich.

It was the French, in particular card manufacturers in Rouen at the end of the fifteenth century, who opted to represent the suits with simple silhouettes of the icons. The need for clearly identifiable silhouettes might also have dictated the ultimate range of shapes we see in the depiction of the clubs and spades. The easy printing of these symbols using stencils, combined with French advances in card manufacture in the fifteenth century, meant that packs could be produced at much lower cost than the hand-painted packs previously in circulation. The French cards flooded the market and soon found their way to England and eventually the rest of world. They were no longer the preserve of the rich but played by all strata of society. This is the reason that spades, hearts, diamonds, and clubs are recognized around the globe, while roses, shields, acorns, and bells remain a niche set of suits confined to the valleys and mountaintops of the Alps.

The French were also responsible for introducing the queen as one of the face cards. In German packs you will find the *obermann* (upper man) and *untermann* (lower man) sitting below the king. This mimics the packs that emerged from the Mamluk sultanate in Egypt, where the king is attended by a *na'ib* (governor) and *thani na'ib* (vice governor). With the introduction of the queen, the British even went as far as reversing the order of superiority during periods when a queen was on the throne.

The original Chinese money cards came with a natural order. Myriads were greater in value than strings, which in turn were greater than the simple coin from which they were all built. This might have led to certain games also adopting an ordering of the suits. For example, in bridge the hierarchy in bidding places spades as the most powerful suit, followed by hearts, diamonds, and finally clubs. However, in other games you will find a different ranking. It appears that the Chinese order didn't carry through as the cards changed and mutated across Europe.

A curious fact: Have you ever looked at the eight of diamonds and spotted that the diamonds are laid out so that the white background reveals a number eight? When someone first showed this to me, I thought they must have manipulated the image. But when I took a selection of packs and examined the eight of diamonds, I got a shock. I'd never noticed that the diamonds actually don't have straight edges. The concave edges result in the figure eight hiding in the card.

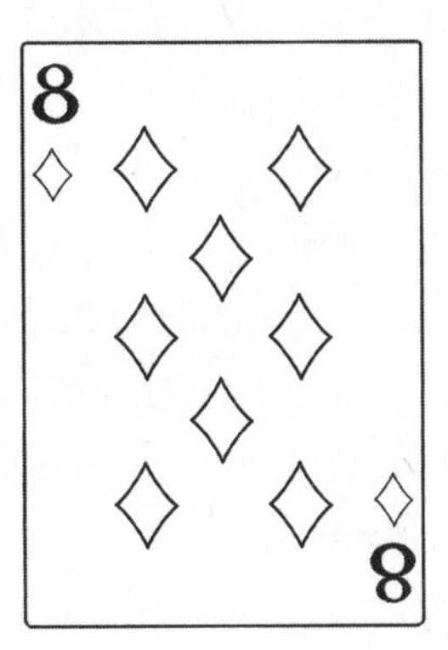

31: Lady Charlotte and the Game of Parliament

MY TRAVELS AROUND THE WORLD have uncovered a fascinating array of different playing cards: the Indian Ganjifa cards, the Chinese Zi

Pai cards—even my son's desire to "catch them all" helped me learn about the more modern phenomenon of Pokémon cards. But collectible cards date back further than we might realize.

The nineteenth century was an age of collecting. Scientists enjoyed classifying, collating, and counting everything they encountered on their travels: rocks, butterflies, fossils, seeds. The craze for collecting and collating went beyond the scientific community, and one of the great collectors of ceramics, fans, and, more intriguingly, playing cards was Lady Charlotte Schreiber.

Her passion for collecting was probably a substitute for the loss of her son Augustus, who died tragically at the age of twenty-one in 1862, possibly from an infection that she had given him. The diaries that she had kept for years go quiet for six years following his death but start up again with her renewed interest in collecting ceramics. Her acquisitions went well beyond just a passing fancy in a market she might have encountered. She treated her collecting as a full-time job with an exhausting schedule that took her from Portugal to Poland, Belgium to Sweden. When her maid accompanied her on a trip one afternoon, she was so shocked by the amount of walking involved that she had to be sent home, while Lady Charlotte checked out the remaining antique houses on her list. The diaries are full of her purchases of china, which would eventually be donated to the Victoria and Albert Museum in London and can be found in ceramics room number 139.

China mania was not uncommon among Victorians, but Lady Charlotte's passion for collecting cards stood out. She admits that in contrast to her knowledge of ceramics, she was somewhat of a novice when it came to collecting cards: "I am so inexperienced as yet, in this branch of collecting that I do not know if they are worth anything," but it soon became a passion that "occupies me very much."

Her collection reveals that playing cards were a vehicle in England not just for gaming but for education and even political messages. One set of cards she collected helped players explore geography. It was believed at the time of the cards' manufacture in the mid-seventeenth century that there were only four continents, which rather neatly matched the four suits in the pack. As the set is French, France occupies the king of hearts, while *les îles britanniques* are relegated to the three of hearts.

Lady Charlotte especially enjoyed cards that had a political message. Her collection reveals the depth of anti-Catholic feeling in Tudor and Stuart England. One pack documents the popish plots from 1558 to 1679; the king of hearts, for example, depicts the plot to murder Charles II, fabricated by the English priest Titus Oates.

An extraordinary pack in her collection documents the frantic stock market speculations that resulted in the crash of 1720. The cards provide political caricatures of the "bubble companies," like the South Sea Company, that collapsed during this period. The ace of clubs depicts the lute-string manufacturers with a caption: "By bubbling too, they've broke their ancient rules, they first made lute strings, but they now make fools." Whether the card is the three of clubs or queen of diamonds certainly plays second fiddle to the illustrations and text that appear on it. She initially was only able to find an incomplete pack of the bubble companies, but her friend Sir A. W. Franks later sourced a complete pack for her.

After she purchased a rather risqué set of cards, she wrote to a friend, "As for the objectionable cards: I have already promised you that I will not even look at them, but as soon as I have made my usual descriptive catalogue of them, I will put them away in a drawer and lock them up, quite out of sight."

She also obtained a curious pack of forty-five cards titled "The Game of Parliament." Each card caricatures a fictional member of parliament with burlesque titles such as "Mr. Charles Badlaw" or "Mr. Partyknell." There are two suits: Liberals and Conservatives. There are cards dedicated to the political programs of each party. For example, the fourth law on the Liberal agenda is "to sell all bachelors over 35 by public auction," while the fifth law on the Conservative ticket mandates "that the sea shall be calm when Lords of the Admiralty sail." The aim of the game is to play your politician cards to try to get all six laws on your ticket passed.

Besides cards touching on British politics, Lady Charlotte also collected a curious French game of cards surrounding the events of the

Revolution. Called "Le Jeu de la Revolution, ou nouveau Jeu mathématique à triangles," it has thirty-three triangular cards that can be placed together to create an image of the Bastille (although Lady Charlotte's set was missing a card).

Despite her coming to collecting quite late in her life, Lady Charlotte's collection is nonetheless impressive and now belongs to the British Museum. So respected was she as a collector that the Worshipful Company of Playing Card Makers, established in 1628, granted her the Freedom of the Company in 1892. She prepared twenty-five volumes of notes documenting her research on the origins of the cards she had in her collection. Like any good card collector, she made trading part of the game. With Franks she swapped a little box of golden ornaments for his collection of 170 packs of cards.

Her collection includes cards she obtained in Italy that collectors praised as "the finest in Europe today." These Italian cards were known as *trionfi* and consisted of a fifth suit of cards with emblematic designs that trumped all other cards. These Italian cards we now know by another name: tarot cards.

32: Tarot

MANY PEOPLE FIRST ENCOUNTER TAROT CARDS in the movies—for instance, in James Bond's *Live and Let Die*, where the protagonist has his fortune told. Others may have a personal encounter with a fortune teller at a country fair or in a new age shop. I first came across the cards when my mother used to visit a lady in Covent Garden who would tell her fortune. The images of the Hanged Man, the Fool, the Tower, and the Wheel of Fortune summon up the mystical power of these cards to divine your future. Some have tried to argue that these iconic cards have origins dating back to ancient Egypt. But the reality is that their use for divination is much more modern than mystics would like to let on.

As I began my research into the true origins of the tarot cards, I was deeply surprised to find that my former colleague at New College,

Professor Michael Dummett, professor of mathematical logic, was the world's premiere expert on tarot cards. I was shocked: Could this really be the same Michael Dummett who was an expert on Gottlob Frege and Kurt Gödel? I had bumped into him regularly as a research student at the logic seminars in the mathematical institute, where he would sit chain-smoking below a very large no-smoking sign, creating so much smoke that his hair was colored yellow from the tar. I knew that Dummett was interested in more than esoteric mathematical logic as he was very politically active, especially in his campaigning against racism. But his obsession with tracing tarot cards through the centuries also reassured me that my own gaming obsession was not a complete contradiction of my pure mathematical sensibilities. "A game," he wrote, "may be as integral to a culture, as true an object of aesthetic appreciation, as admirable a product of human creativity as a folk art or a style of music; and, as such, it is quite as worthy of study."

It seems that Dummett's political activism may have been the catalyst for his dedicating time to tarot. During a particularly stressful political period, he found it difficult to concentrate on the intellectual demands of pure philosophy and discovered that researching a more historical topic offered an alternative to the political battles that were raging.

Dummett first came across the tarot cards as a child when he was given a book on cartomancy. The book mentioned that in addition to their role in fortune telling, the cards were also used for a game. "This piece of information stuck in my mind," he said. "Like many others, I was fascinated by the Tarot pack and, though I had no belief in the capacity by its means to foretell the future, I was consumed with curiosity as to what sort of game could be played with it."

A tarot pack consists of four suits corresponding to cups, swords, coins, and batons. Each of these suits consists of fourteen cards, ten numbered and four court cards: the king, queen, knight, and jack. But it is the additional fifth suit that is so intriguing, consisting of twenty-two special tarot cards numbered using Roman numerals. The Hanged Man, for example, has number XII. These beautifully illustrated cards make up the Major Arcana and often play an important role in fortune telling. But was that why they were introduced, and what were the cultural origins of this strange collection of characters? Dummett believed that fortune telling was a red herring and that understanding how to play the games with these decks was the real key to revealing the origins of the Hanged Man or the Wizard.

On his European travels, Dummett sought out bars and cafés that were still using these cards and would insist on joining in to try his hand at the local game variety. If you invited him to a mathematics conference, then invariably he would put in a request for the local organizers to search the bars for a local game.

In his seminal work *The Game of Tarot: From Ferrara to Salt Lake City*, a six-hundred-page epic, Dummett documents references by the church and state to the evil nature of tarot, never in relation to magical powers or fortune telling but always in connection with games and recreation and, more importantly, gambling. At the end of the sixteenth century, Henry III of France taxed all sorts of cards: "The games of cards, Tarot and dice... instead of serving for pleasure and recreation in accordance with the intention of those who invented them, at the present time serve only to do harm and give rise to public scandal."

Dummett argues that the absence of any mention of the occult in the earliest references to tarot demonstrates that the cards have nothing to do with magic. One can really see the mathematical logician's mind at work in Dummett's investigations of whether the occult has anything to do with the origins of tarot and the extra Major Arcana cards. He is applying the same principle that he applies to his philosophical study of Frege or John Locke.

Words only gain meaning when put in context, not when investigated in isolation, or, as Ludwig Wittgenstein advocated, once they are used in the language game. And so the tarot can be considered not in isolation as a mystical set of cards but only in terms of how they were used. As Dummett writes, "To understand the purpose for which the Tarot pack was invented we have therefore to ask for what reason an ordered sequence of cards, of different length and composition from the ordinary suits, was added to the regular pack.... Obviously to find an answer, we have to look at the role that these cards play in the game, on the reasonable assumption that the essential features of the game, in the various forms in which it was later played, belonged to it from the start." The challenge for Dummett was to work back from the games played in the present in the cafés of northern Italy to try to deduce what games might have been played in the fifteenth century when these cards were first introduced. The key seems to be that this is the period in the history of cards when the idea of a suit that "trumps" all other suits emerged in game play. The trick-taking games that arrived in Europe around this time do not include this feature. It seems to be a genuine new innovation in games. Today we are used to one of the four suits assuming this role in a game. But another solution would have been to introduce an independent fifth suit that would always act as trump cards, beating all the others. Indeed this fifth suit was really only called the Major Arcana once the cards were used in cartomancy. Before the seventeenth century they were known as *trionfi*, or "triumphs," because they triumphed over all other cards. This word evolves into the word *trump* that we use in games today.

Since these cards needed to be recognizably different, their inventors needed to mark out the fifth suit in a way that differed from the other suits. Hence we have the procession of twenty-two allegorical figures that would have been recognizable to Renaissance courts, where alchemy, astrology, and divination were certainly present. But in other sets of tarot cards from Austria, Bavaria, and Belgium, we find the fifth suit being represented by animals and plants, with no connection to the occult.

Dummett had no interest in the mystical associations of tarot cards. Evidence now suggests that the fad for using these cards for divination only started in the 1700s—not that Dummett cared. Woe betide anyone who tried to engage him with the divinatory role that the tarot has played in Western culture—pure propaganda peddled by the occultists. For Dummett, it was the game that mattered; the rest was complete and utter nonsense.

You could spend a lifetime collecting the different varieties of tarot cards that have been published over the centuries, but we are moving on to discover a whole host of different cards to collect. Phileas Fogg's steamer from China across the Pacific was not direct to America but went via Japan, docking at Yokohama. So before I head on my journey across the Pacific Ocean, I shall also make a detour via Japan to discover an array of interesting playing cards that the country enjoys playing with.

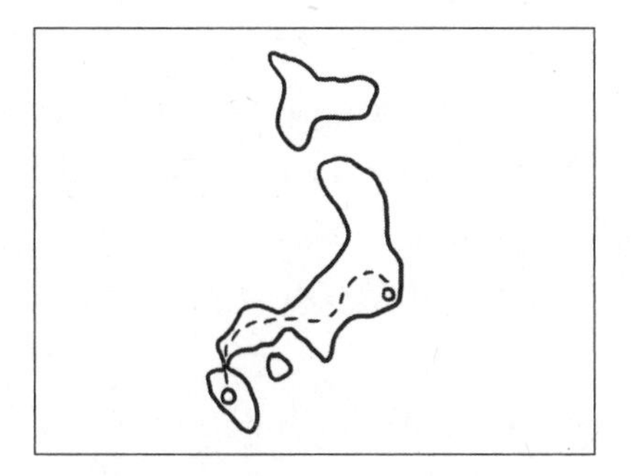

CHAPTER 7

JAPAN

JAPAN ENJOYS MANY of the games of its neighboring countries, games like Go and chess, but it is also home to its very own idiosyncratic games, as I discovered on my mathematical visits to work with my Japanese collaborator. These include a card game that I've never seen outside Japan and another card game that has taken over the world.

33: Hanafuda

"PEOPLE DIVIDE INTO TWO CAMPS. Those who like their desserts and those that don't. Professor du Sautoy San, which are you?" This was the first thing that Professor Kurokawa asked me as we met in the arrivals hall of Haneda Airport in Tokyo. I'd already had rather a nerve-racking time trying to locate my host. Upon my arrival, it suddenly occurred to me that I had no details of where I was heading once in Japan. No one was holding up my name, and I was feeling somewhat lost when I noticed a man in a bucket hat wandering around with a sign with a huge Greek letter denoting zeta on it.

It was then that I knew I'd found my host. We are both experts in the mathematics of zeta functions, and our shared passion is what had brought me to Japan. Judging from Professor Kurokawa's waistline, I was guessing we also shared a love for desserts. He was very happy to hear that we could begin our collaboration by heading straight to a café serving matcha-flavored desserts of many different green hues.

Professor Kurokawa was the most wonderful host. Once he knew I was obsessed with the mathematics of symmetry, he took me up to the beautiful Shinto shrines and Buddhist temples in Nikko. When I mentioned my passion for theater, he rewarded me with a visit to the kabuki theater. This incredible experience involved the audience shouting out the names and numbers of the actors as they came on stage as if we were chanting for our favorite football players.

Best of all, when I mentioned an obsession with collecting games as I made my travels around the world, he took me to a shop to track down a set of cards that are particular to Japan, called *hanafuda*, or "flower," cards. Like everything in Japan, the design of these cards is a thing of beauty. They are much tinier than any cards I'd come across before, around 3.5 × 5.5 centimeters. They were also much thicker, with a pack of forty-eight cards sitting more than twice as high as a classic pack of fifty-two Western cards.

Playing cards have a long history in Japan. First introduced into Japan by the Portuguese in the mid-sixteenth century, they were banned when Japan closed its doors in the 1630s to the influence of Western traders, believing that foreign influence was corrupting Japanese culture. Having already got a taste for gambling with cards, a cat-and-mouse game ensued between the authorities trying to clamp down on the production of cards and manufacturers trying to produce cards that might evade the bans. This might be one reason that the cards ended up being so small: they could quickly be hidden in your hand if the authorities happened to pass by the café you were playing in.

This also might explain why I'd never seen anything like the Hanafuda cards before. They are divided into twelve groups of four cards. Each group corresponds to a month of the year. (There is evidence that at one stage the pack consisted of one hundred suits, but this was cut down because a pack of four hundred cards was too unwieldly for gaming.) Each month has a flower associated with it. The beautiful Japanese cherry blossom is the flower of March. The chrysanthemum is September's flower, while October features the stunning colors of the maple leaves as they turn in the autumn.

Though the cards may have been recognizable to a Western audience in the sixteenth century when they were introduced, it seems that designers tried to make the illustrations as different as possible to further distance them from the kinds of Western cultural artifacts authorities were trying to ban. Hence we see no numbers or pips, just cards with these beautiful pictures of flowers. Some of the cards include additional imagery, which gives them increased value in games played with the cards.

The most valuable cards are the *hikari*, or "bright," cards. These consist of a red-crowned crane and sun (from the January suit), the camp curtain (from March), the full moon (August), the rain man carrying an umbrella (November), and the phoenix (December). The moon of August is probably the most iconic of all the cards and is often used as the poster card for the whole set.

The next most valuable are the animal cards, although not all of them are actually animals. These include a Japanese warbling white-eye (from February), a lesser cuckoo (April), an eight-plank bridge (May), butterflies (June), a Japanese boar (July), geese (August), a sake cup (September), a sika deer (October), and a barn swallow (November). August and November have both bright and animal cards in the set. The third category of cards with extra value are the ribbon cards, followed by the simple flower cards without any extra ornamentation.

A classic game played with these cards is Koikoi, which is frequently featured in Japanese films. It is a matching game where you claim cards from the table by matching them with cards from the same month in your hand. You then try to piece together special collections of cards called *yaku*, which score a variety of different points. For example, collecting all the bright cards scores you fifteen points, while getting the boar, deer, and butterfly will score you five points. When you make a yaku, you can call for the end of the game, but your opponent can reply with *koikoi*, meaning "come on!" in Japanese, and the game continues. This is a risky call, because if you can't make a superior yaku, and your opponent succeeds in extending their yaku, they score double points.

Hanafuda cards have played a crucial role in the history of gaming by helping to kick-start what became one of the biggest video game manufacturers in the world. It all began when craftsman Fusajiro Yamauchi decided to open a card shop in Kyoto in September 1889 to sell Hanafuda cards painted on mulberry tree bark. He called the company that he founded Nintendo Karuta.

The Nintendo cards were very popular up until the 1960s, when they fell out of favor as they began to be associated with organized crime. The Japanese gangsters known as the Yakusa were fond of Hanafuda, and many gang members would have tattoos made of their favorite cards. Just as Hanafuda cards fell out of favor, the Nintendo company saw the advent of computer games in the 1970s as a possible new direction for the company. So it was that one of the biggest game manufacturers in the world can trace its origins to the Hanafuda cards that were sold in the side street shop in Kyoto that Yamauchi opened in 1889.

34: Pokémon Cards

ALTHOUGH THE HANAFUDA CARDS that helped create Nintendo may not be known the world over, Nintendo did have a hand in marketing another set of gaming cards that have taken the world by storm: Pocket Monster cards, better known as Pokémon for short. The first generation consisted of 102 cards that players would collect, trade, and compete with. They arrived in the United States in 1998, and four hundred thousand packs, each containing ten cards, had been sold in less than six weeks of their release. The cards arrived a year later in the United Kingdom, and it wasn't long before my son and I were pitching Charizard cards against Squirtles and going down to the local newsagent to buy yet another pack in the hunt for a Mewtwo.

The idea behind the game is that each card represents a character with its own strengths and weaknesses. Players choose a card to represent them in battle and then try to use the character's abilities to inflict damage on the opponent's card. The aim is to reduce the hit points of the card to zero and knock it out of the game. The game relies on a principle called nontransitivity to create a deck that is well balanced. As in Rock Paper Scissors, it is important that each card be strong against some cards even as they remain vulnerable against others.

The popularity of these cards was partly due to the combination of trying to collect all 102 cards to complete the set, together with the fact that there was a game you could play against friends. When was I collecting football cards as a kid, no one had yet had the clever idea of using the cards to create a game.

The combination of collecting, trading, and gaming was first introduced with a set of cards called Magic: The Gathering in 1993. In this game you are a wizard pitching your spells and magical creatures, represented by the cards in your hand, against your opposing wizard. It has something of the flavor of Dungeons & Dragons. Indeed, its creator, mathematician Richard Garfield, was inspired to create the game after hearing about Dungeons & Dragons but being unable to find the game in his local store.

Games had always been important to Garfield. He spent some of his childhood in Bangladesh and Nepal. Unable to speak the languages, he was able to connect with local kids through playing games like marbles. Since he couldn't get his hands on this exciting new game of Dungeons & Dragons, he decided instead to create his own fantasy game.

The game play involves a good deal of strategy in choosing the cards you will fight with. But the element of luck involved in drawing cards from your pack means that a beginner can still have a chance against an experienced player.

Garfield was keen to ensure that games of Magic: The Gathering involved some jeopardy, and so in the early days players would have to randomly stake a card from their collection as potential payment for losing a game. Called the ante, the rule actually ran into trouble in territories where gambling is forbidden.

The game that Garfield created has a rather extraordinary property. It was recently proved that the game is sufficiently complex that it is possible to encode a universal computer into the cards, which means that any algorithm that can be implemented by a computer can actually be simulated in the game play. For example, since you can write an algorithm to find the millionth prime number, there is a play of cards that can be interpreted as doing the same calculation. This is done by arranging a series of cascading cards that trigger each other in such a way that each player's actions are forced by the sequence of cards. It's a highly unlikely scenario but not forbidden by the rules of the game.

Since it was proved that it is undecidable whether an algorithm implemented on a universal computer will actually halt, the implication is that there are scenarios where it is impossible to determine who will win the game. No algorithm exists that can determine whether in certain scenarios there is a winning move—or, even more strikingly, whether the game will terminate at all.

The reason that cards can be used to simulate the way a computer works is due to the nature of the instructions to be carried out when playing the card. For example, the Xathrid Necromancer card can be used as a component in the computer because it instructs the player "Whenever Xathrid Necromancer or another Human creature you control dies, create a tapped $\frac{2}{2}$ black Zombie creature token." This effectively acts like a line of code: "If the machine is in state *s*, and the last read cell is symbol *k*, then do such and such."

Magic: The Gathering was an almost immediate success when it was released, with the ten million cards that had been printed selling out within two months of release. The company creating the cards had to stop advertising the game to keep pace with demand.

Pokémon cards built on the advances of Magic: The Gathering, using easier rules and accessible graphics geared to the younger crowd to create a kinder, gentler version of a collectable, competitive game. Like Magic: The Gathering, Pokémon too became a phenomenon.

Pokémon cards recently went through a resurgence of interest as rare cards from the first generation began fetching crazy prices in auctions across the world. However, cards only had value if they'd never been played with. This sad fact is reminiscent of the ant in Aesop's fable, who squirrels away his cards for the future, unlike the grasshopper, who enjoys playing with them in the present. The most expensive Pokémon card ever sold was a 1999 Pokémon Base 1st Edition Holo "Thick Stamp" Shadowless Charizard. It sold for $350,100 on eBay on December 12, 2020. With prices like that, it may be worth digging out my son's old collection of cards to see whether I might be sitting on a goldmine of old Pokémon after all.

With the development of the metaverse and blockchain technology, trading cards have moved from the analog to the digital realm. The blockchain has provided the platform to identify the unique owner of a digital trading card via the technology of nonfungible tokens (NFTs). The digital image might be copied many times, but the NFT allows one person to own the "original." Just as collecting cards evolved into a game based on your collection, NFT gaming is beginning to catch on as well. Take Axie Infinity. Axies are virtual monster-like pets that players can create on top of the blockchain, which they can then battle and breed with other Axies owned by other players. They are, in effect, NFTs crossed with Pokémon cards.

Axie Infinity is just the beginning of what may ultimately make up a whole new world of games that will emerge in Web 3.0 and the metaverse. This may well be the future of games—but to understand the full scope of the gaming universe, we are going to take a trip during our next sea interlude into the human mind to understand something about our compulsion as a species for playing games and what impact it has on our brain.

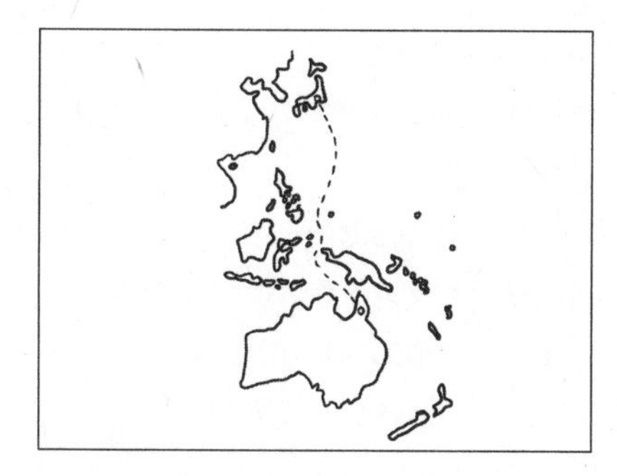

CHAPTER 8

THE NORTH PACIFIC OCEAN

The Psychology of Games

THE PLAYING OF GAMES serves multiple roles in our psychology. As we've seen, games provide us with that dopamine hit that makes us feel good and a safe environment in which to get to know people. There is a close connection between games and ritual. The following of rules in both provides a way of escaping time and place. They help one to transcend the worldly.

Psychologists Edward Deci and Richard Ryan suggested in a seminal work published in the 1980s that human behavior is driven by three needs: competence, autonomy, and relatedness. Games provide all three. The state of mind that we often seek to achieve while playing a game is something psychologist Mihály Csíkszentmihályi calls a state of flow, where challenge and ability are matched so that we get lost in the moment of play. A game is entered into freely; it is an assertion of our free will and plays to our agency in the world. This satisfies that need for autonomy. But a game is also about playing with others and sharing a space and time, satisfying our need for relatedness.

A game provides a space to be the person we would like to be or to try out a new personality. That we enjoy playing games has been exploited as a way to incentivize us to do activities that without the gaming element we could easily give up. The buzzword is *gamification.*

Everything is being turned into a game.

But games can also carry a dark side. Getting lost in a game can be seen as losing touch with reality, reneging on our responsibilities in the world, and even corrupting the way we think. A game can become so addictive that it becomes unhealthy. The dopamine hit of winning at the roulette wheel can lead someone to become bankrupt trying to chase that hit. Children in China have had to go into detox clinics because they have become so addicted to the world of the online game. Perhaps this is why Buddha banned so many games in his time.

To begin to assess the psychology of games, we will begin this sea interlude with a game that has been celebrated as a fantastic forum for providing a safe space for the neurodiverse to explore aspects of their psychology that the real world might have difficulty accommodating. It has also been regarded as so addictive and corrupting that headlines in the 1980s declared that it was nurturing a satanic cult indulging in ritualistic killings.

35: Dungeons & Dragons

CHRISTMAS 1983. I was eighteen years old. Spotty. A nerd. Big into games. A present sat under the tree addressed to the family. My mum handed it to me and my sister to open. As the paper ripped apart, the image of a dragon emerged sitting atop a pile of gold. A wizard and a knight were about to engage it in battle. It looked amazing. The box declared, "You are holding a fantastic world of swords and sorcery adventures in your hands! In Dungeons & Dragons you become a mighty wizard, a fearless hero, a stout dwarf, a clever halfling or any one of a dozen other adventurers ready to explore the mazes and labyrinths of a vast and deep dungeon."

Wow! Here was my chance to live out my fantasy of being in a Tolkien story. We opened the box, excited to arrange all the pieces on the board. Except there wasn't too much inside—just dice and a lot of rules. But what dice!

In addition to the classic cube-shaped dice of our Monopoly board, there were some of the other dice that I'd already understood were available since my forays into ancient Greek mathematics. A twenty-sided icosahedral die. A four-faced tetrahedron. A die made of twelve pentagonal faces called the dodecahedron. And finally the eight-faced octahedron. Someone had obviously done their research. One of each of the Platonic solids. I was already intrigued. But apart from the dice, there was just a lot of reading material that started to feel worryingly like an examination booklet. "You may turn over your papers now."

Someone had to volunteer to take this exam in order to become Dungeon Master. The role seemed to require becoming a novelist and creating the setting and scenarios that the rest of the family would encounter as they descended into the dungeon. It all seemed a bit daunting until I found that part of the exam material consisted of a pre-prepared map of a two-level dungeon, plus a sort of script to help me narrate players' journey through the dungeon. It would be my job as Dungeon Master essentially to populate the dungeon with a selection of monsters and hidden treasure. The gauntlet was thrown down. I accepted the challenge.

I spent several days that Christmas in my room studying for my exam to become Dungeon Master. Eventually I felt ready to take the rest of the family on their first adventure: In Search of the Unknown. One of the unique things about this game is that rather than commanding an army that you push around the board, like in chess, you assume an individual role. You can choose between becoming a wizard or a dwarf or a knight and then act out the adventure through your decisions in the team. This was one of the first of what would become known as role-playing games.

Also unique was the way the game allowed players to work together rather than against each other. Although the idea of collaborative board games wouldn't formally appear on the scene until the early twenty-first century, Dungeons & Dragons exhibited a shift in mentality. Later collaborative games like Pandemic have clear goals, and you either win as a group or the game beats you. That is less obvious in Dungeons & Dragons. It is more like the kinds of open games or infinite games where the task is more about keeping the narrative going. There might be a dragon guarding a treasure hoard whose conquest marks a successful mission, but this game has the flavor more of reading a novel where the last page is a very small part of the whole experience. It's about how you get to that denouement.

Once my dungeon was ready, the family had to choose their roles. Like the hand shapes in Rock Paper Scissors, characters are strong in some realms but weak in others. Fighters have strength but lack the intelligence of the magic users. You need to put together a team that is balanced in the six key traits possessed by each character: strength, intelligence, wisdom, constitution, dexterity, and charisma. Every challenge the team encounters in the dungeon will favor different characters on the team.

Dad chose to be a human fighter named Brakspear of Uist. We grew up in Oxfordshire, where the local beer was brewed by Brakspear. Our summer holiday that year had been in the Scottish islands. My mum was clearly going to go for being a wizard, given her interests in the occult: Marok of Cornwall. To balance the team, my sister assumed the role of Shortley, the Farlogian Thief, a dwarf. The dice were rolled to determine the different values of the characters' attributes, and then money was spent on armor and provisions in preparation for the trip. I've still got my sister's inventory that she chose that Christmas: kinky black leather armor at a cost of fifteen pieces of gold, smoked salmon and avocado for another fifteen pieces of gold (useful for distracting hipster hobgoblins), and a hip flask of Southern Comfort. This dwarf was ready to party.

With everyone kitted up, the family set off on their adventure. As they made their way through the dungeon, my script helped me to describe each new room or cavern they encountered. A piece of graph paper snatched from an old math exercise book allowed them to map out the contours of the emerging dungeon. Monsters they met on the way were battled using the throws of the dice. Each new room offered choices about where to go next.

The mechanism actually shares a lot in common with a whole genre of "choose your own adventure" books. Turn to page thirty-three if you want to descend the staircase or page forty-seven if you want to try the door in the far wall. "Bandersnatch," a recent episode of *Black Mirror*, even offered a TV version of the same idea. Capitalizing on the success of Dungeons & Dragons, Ian Livingstone and Steve Jackson published a whole series of adventure books that you could play on your own, rolling dice and choosing options as you explored the adventure. Called Fighting Fantasies, they were extremely popular. In March 1983 the top three entries in the *Sunday Times* best-seller list were Fighting Fantasy titles. They sold twenty million copies in the 1980s and 1990s.

The game of Dungeons & Dragons, or D&D as it's fondly known, became a phenomenon of the 1970s and 1980s. It's why the cult Netflix series *Stranger Things* uses it to immediately take audiences back to the 1980s. But as computer games became ever more sophisticated, the power of code to navigate your way through the multiple choices involved in exploring a fantasy world took over from the need for a human to do all the work of preparing the world in advance. Yet, powerful as such fantasy worlds as Elden Ring, Tomb Raider, or the Legend of Zelda may be, Dungeons & Dragons has seen incredible growth in the last few years. Indeed, 2020 was the game's most successful year since it first was launched in 1974, with over fifty million players worldwide.

What is the appeal of this analog version of an adventure that many have got used to playing online? Some have suggested the pandemic as a reason for its recent success. Suddenly personal contact with a group of friends playing a game around a table has become sought after in a way that online gaming with friends can't replicate. We've had enough of the virtual world of Zoom. But that doesn't explain why the game has grown in popularity year on year since the publication of the fifth edition in 2014, long before the pandemic.

Others have suggested that featuring the game in a central role in the cult hit series *Stranger Things*, launched in 2016, has helped boost its popularity. The characters use the game as a way to navigate and articulate the strange occurrences that are happening in their real world. In contrast to the stereotypical nerdy, white, young male gamer—me—who was attracted to D&D during the 1980s, the recent surge in popularity has seen a much more diverse community being drawn to the game.

People have acknowledged how powerful an experience it is to assume a character with diverse traits and be accepted for who you are. It can offer a safe space for exploring part of your personality that you are nervous about expressing in a public social environment. With people exploring different gender identities, the possibility of playing a different gender in the environment of a game has the potential to be super liberating. My experience of assuming a female character called Zeta Siegel in the virtual world Second Life gave me unexpected insight into the sexual harassment women encounter. I still feel a little guilty about leaving Zeta frozen in the metaverse after I stopped exploring the thrills of this virtual space.

Introverts especially have found Dungeons & Dragons a very safe space to take the risk of being more extroverted. This perhaps is particularly relevant to those on the autism spectrum. It is a controlled environment, with rules and a time limit, which means you know when it will end. The game sends a message that diversity is a superpower, a sentiment Greta Thunberg promoted with the hashtag #aspiepower in connection with her diagnosis of Asperger's syndrome.

Diversity is key to battling through the dungeons because you need people with complementary skills. If you're all the same, your campaign is unlikely to succeed.

The internet has provided a space for hugely elaborate expansions of the concept of Dungeons & Dragons. Games like Dark Summoner and Villagers & Heroes allow players to create characters and join clans to go on virtual quests. My mother has found these worlds a fantastic escape in her old age, where she can assume the role of a great warrior and not be judged as a woman in her eighties. I expect my inheritance might include a rather splendid array of virtual armory that she has accumulated over the years. I discovered recently that thanks to her status in these games, she'd found herself a key member of a clan whose other members in real life turned out to belong to a Mexican drug cartel in Chicago and to enjoy playing the game when not dealing amphetamines. Amazing how a game can provide a common space for such diverse individuals to interact. These games all provide that sense of relatedness that our psychology craves.

Ultimately perhaps board games and role-playing games like D&D tap into the very human need to sit around a campfire and tell stories to each other. In an age when we are getting tired of algorithms pushing us around, the computer game just feels a bit too close to the controlling nature of modern code, picking our movies or choosing what we listen to. We are hankering for something to complement our increasingly digital existence, where our work, social, and love lives are increasingly dictated by our online presence. It could be that the performative nature of social media has stimulated an appetite for playing out roles and trying on new skins and that this is tipping over into finding an analog platform to continue this exploration.

Our own family experience of D&D didn't really survive that Christmas of 1983. There was a moment when the gang encountered a goblin I'd planted in the lower realms of the dungeon. My dad's fighter was having trouble battling and was close to being killed. My mum cast a sleeping spell, but then they decided to kill the goblin while it was asleep. But after they'd done this and claimed the treasure and hit points that were awarded for the kill, there was a collective sense of unease. Should we really have killed a beast, however scary, while it was asleep? We left the game and never came back to it. I think it had revealed something about us that we'd actually have preferred not to discover. The box was relegated to the cellar, where it has sat ever since.

36: MangaHigh.com

GAMES HAVE A DEMONSTRABLE power to engage people in hours of activity, trying to clear a new level of a computer game, plan a sequence of moves to achieve checkmate, or build up a property empire to defeat opponents in Monopoly, among many other time-consuming tasks. For some, this power to incentivize behavior, if only it could be harnessed, promises to achieve extraordinary things. The idea has even spawned its own word: *gamification*.

Today everything is being gamified. Your fitness training is changed into the challenge of beating your friends to the top of the leaderboard. Learning a language is rewarded by medals for lessons completed. In the workspace, turning mundane tasks into part of an ongoing quest gives a story to work that might otherwise become monotonous. Health apps incentivize engagement by allowing you to pimp out your avatar with new accessories that you can show off to friends who use the same app. The key to all gamification is to keep those three human psychological needs in mind: competence (it should be challenging but not impossible); autonomy (you must feel like you are making the decision to engage with the game); and relatedness (it exploits the desire to share your achievements with your social group).

Given how much kids in particular like gaming, education has been a particularly fertile space for gamification. I have myself indulged in a bout of gamification in an effort to teach students mathematics. Toby Rowland, a friend from university, and I collaborated to create the first online math school based on gaming called MangaHigh.com. Toby had already made his fortune creating a site called King.com that tapped into people's love of casual gaming. Players paid a fee to enter tournaments or play games head-to-head, with the highest-ranking players winning part of the money that players had staked. One of the platform's games, Candy Crush Saga, went on to become one of the most financially lucrative games of all time. The success of the game can be judged by the fact that on every trip I've made on the London Underground, I've sat next to someone playing the game.

Toby was struck by how many hours players would invest in trying to upgrade their status in the game, clear levels, and beat their fellow gamers. The amount of skill necessary to achieve their goals was extraordinary. And yet these hard-won skills were totally useless outside the game. Toby had a vision: What if the skills that someone learned to clear level after level might actually be useful in the outside world?

His question to me was whether we could use games to teach mathematics.

It was a fascinating challenge. Of all the subjects in the educational curriculum, mathematics did indeed seem perfectly suited to making into a game. It played into my whole thesis that mathematics and games are intimately related. Just as you know when you've won a game, mathematics has that quality of having a right answer or not.

Toby enlisted two big names that knew how to create addictive and successful games. German game designer Reiner Knizia is famous in the gaming world for designing over seven hundred games, many of which have won international prizes. As if to confirm my belief that mathematics is the key to creating a good game, I discovered when we first met that Knizia had originally studied mathematics at university. The second big name in game design that came on board was Ian Livingstone, the man behind the gaming books that tapped into the Dungeons & Dragons craze. Livingstone is a big believer in the power of games in education and was asked by the UK government to advise on how games might play a role in schools.

The dream team assembled, we began to turn the whole of the school curriculum, from learning your multiplication tables to solving quadratic equations, into a game. The criteria for a good game were strict. It wasn't good enough to simply challenge kids with a bit of math to solve to gain a reward or badge or to move up a level in the game. The mathematics that we wanted the game to teach had to be integrated as a fundamental part of the game play, so much so that by playing the game you were experiencing the power of the mathematics to help you achieve your goal. Step by step each piece of the curriculum turned into a game.

Factorizing quadratic equations became Wrecks Factor, a game where you rescued ships by creating rectangles around the wrecks whose sides corresponded to the factors. The game forced you to stop thinking of quadratic equations as abstract algebraic notations and conceptualize them as something more concrete, like actual areas of land. Tangled Web required players to use the mathematics of angles to solve puzzles that built the web to help Itzi the Spider to climb through the levels and rescue his family. In Pyramid Panic you've got to help a prematurely entombed mummy escape from a pyramid by solving geometry puzzles to build a path across the burial chamber to the exit. Algebra Meltdown capitalizes on the fact that many algebraic equations are like cryptic crosswords where the equation is the clue hiding the identity of the unknown x. They are perfect for embedding at the heart a game that actually takes you through the process of undoing this puzzle.

In Save Our Dumb Planet, the player has to protect Earth from meteors and other space hazards by calculating trajectories for a life-saving surface-to-space missile. Players use algebraic substitution, indices, coordinates, and graph plotting to plan their missile flight paths, leading them through linear, quadratic, and eventually cubic equations. It is not dissimilar to the actual skills used by real-life scientists at NASA to steer space crafts through the solar system.

And so MangaHigh.com, our internet math school, was born, with the mission of turning the whole school curriculum into a game. Given that Japanese manga is often set in a school context, we chose to illustrate our school in the style of a manga comic. The games in MangaHigh act like teachers. Because the mathematics is an integral part of the game play, it isn't just testing your understanding; one can actually learn the mathematics by clearing levels. The best educational games are those where you learn as you play, where you don't need to know anything before you play, but as you play your experiments help you to understand the mathematical ideas that are at the heart of the game.

For me, math has always felt like a game. You get stuck on a level. You try out different strategies, and then you get that buzz of satisfaction as you crack the challenge. We had a lot of fun creating the games behind MangaHigh, finding ways to make the mathematics curriculum into exciting, fun, but challenging games. The true test of whether we were successful though was what kids thought. The reward was witnessing how the games we created engaged kids in the classroom who never previously got excited by math.

And sure enough, after playing our games, they were actually asking for more math homework. Gamification really can work.

37: Cranium

THE GAMIFICATION OF EDUCATION ASIDE, there are also games that are directly pitched as a means for improving brain function. That playing games has a good effect on the brain was one of the big selling points for Cranium. Released in 1998 it was unusual in that it was really a compendium of different games woven into a single game. The creators of Cranium wanted to give the brain a full workout by challenging it and players with a whole range of different activities. Players had to draw pictures, solve puzzles, act out scenarios, hum songs, spell backward, and sculpt models. I think this is the only game I've seen that comes with a tub of clay among its equipment necessary for play. The idea was that players should be able to find one of these activities that they could excel at, making everyone feel like a winner.

The game proved a huge success. It won a number of "game of the year" awards, and gaming company Hasbro gave it its seal approval in 2008 when it bought Cranium for $77.5 million and marketed it alongside its big hitters Monopoly, Cluedo, and Scrabble. One of the big selling points was the idea that it was a "game for the whole brain," and it was one of a whole slew of games that marketed themselves as gyms for the mind.

But does playing games really have much of an impact on the brain? Research has indeed revealed that grandmasters in board games like chess or Go have structurally different brains. fMRI scans showed an increase in gray matter in Go players in the nucleus accumbens, a region responsible for releasing dopamine to aid the brain in processing external stimuli related to situations that are either positive or negative. Such a part of the brain is probably being exercised constantly in assessing the changing nature of the stones on the board. In contrast, the amygdala, an area of the brain responsible for our emotional state, saw a decrease in gray matter. Emotions can often get in the way of playing effectively, so Go players have trained their brains to regulate emotions, leading to a decrease in this region.

There is also evidence of a different kind of connectivity at work in the brain of grandmasters of chess. While an amateur player uses a lot of the medial temporal lobe at the center of the brain to consciously analyze the pieces on the board, a grandmaster's brain bypasses this area. There is instead increased activity in the frontal and parietal cortices, areas more associated with intuition and long-term memory. In a strange way the brain is using fewer neurons to tap into the best move. The grandmaster knows a move feels right without being able to articulate why. The brain doesn't work hard producing a logical rationale for the feeling, wasting energy in the medial temporal lobe.

Some have feared that games, especially video games, might actually have a detrimental effect on long-term players. Headlines often declare that first-person shooter games like Call of Duty lead to more violent individuals. Stories tell of kids in China so addicted to playing League of Legends that they wore nappies in order to avoid having to abandon their screens. China now has boot camps to detox kids overwhelmed by the drugs their brains produce as they play their online games for forty hours at a time.

Research has confirmed that video games do change brain structure. But the surprise for many is that the changes are mostly for the good.

Nongamers who played a first-person shooting game called Medal of Honor for an hour a day found that they were able to focus on tasks with multiple distractions far better than those who were given a more passive video game like Tetris to play. It also increased the ability of female participants to manipulate 3-D objects in their heads, something that male brains are thought to be better at. And the evidence is that first-person shooter games are not responsible for making players more violent in their nongaming lives.

There is research that shows that playing analog games like cards or board games in later life actually increases your longevity and staves off dementia. Games often test memory.

Card games are won by those who memorize the cards that have already been played. Playing games is going to give your memory cells a good workout. A long-term study carried out in the southwest of France on 3,675 elderly participants found that the risk of dementia decreased by 15 percent in those who played board games regularly, and the rate of depression was also much reduced. Games are played together. Company in old age is essential to mental and physical health. A win might give you a hit of dopamine, but playing in a group will release another important feel-good drug that the body specializes in: oxytocin.

Although there are a lot of dark fears about the effects of certain games on the brain, mostly the science seems to show a positive impact. As our journey drops us at our next destination, Australia, we are about to work those neurons and pathways once again.

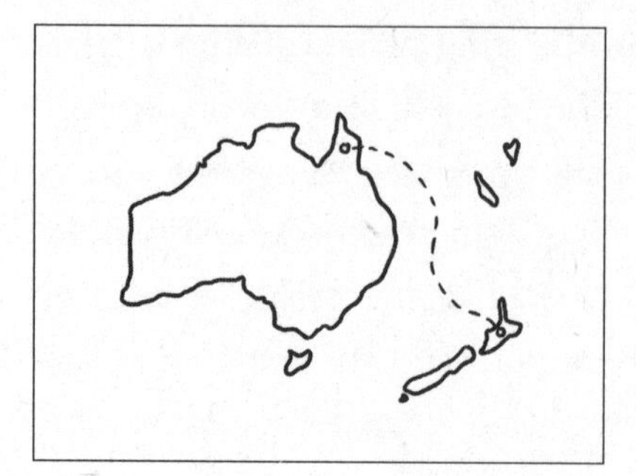

CHAPTER 9

AUSTRALASIA

A FULL QUARTER OF PHILEAS FOGG'S journey around the world is spent on the *General Grant*, the steamer that takes him from Japan to America. But our route will be a more circuitous one, allowing us to visit some important regions that would have slowed Fogg down.

I've visited New Zealand on numerous occasions for work, and over the course of my search for the activities that locals enjoy to pass the time, I discovered that New Zealanders seem to have a particular fondness for anything that involves adrenaline. Bungy jumping. White-water rafting down class-five rapids. Ziplining. On my last trip I bought a T-shirt with the logo "Addicted to adrenaline: the only legal drug."

I'm actually not much of an adrenaline junky myself, so on one weekend I opted for what I thought looked like a gentle stroll through the local nature, something the travel agent called *canyoning*. Little did I know that canyoning would actually involve rappelling down 150-meter waterfalls, ziplining across canyons, and, worst of all, jumping into pools of water from cliffs so high that you had enough time to scream twice before you hit the water. I spent the day ignoring my brain's instructions to stop putting my life in danger. I think by the end of the day, my body had exhausted the adrenaline supplies it was able to generate. That evening I attended a party in the eleventh-floor flat of my collaborator in Auckland. I had to stop myself from going to his balcony in case I'd reprogrammed my brain to think I was supposed to launch myself over the side.

Setting extreme sports aside, most of the games I found on my visits to Australia and New Zealand were invasive species from abroad that probably damaged the local gaming environment as much as the introduction of rabbits and cane toads had harmed the ecosystem. It is a familiar story: the Europeans who came to the region in the eighteenth and nineteenth centuries brought the games they loved playing, and those familiar games ended up defining the gaming culture up to the present. I really wanted to know what games were played by indigenous people.

As I found out, the majority were what I would classify as sports rather than the strategy games that I am after. One example was Edor, a team game in which one player runs with a ball trying to avoid the players on the opposing team. It originates in the Aurukun Aboriginal community in North Queensland in Australia and can be seen as an early forerunner of rugby. Another game along these lines is Kee'an, which simply means "to play" in the Wik-Mungkan language of North Queensland. In Kee'an twine is attached to a large bone such as an emu shinbone, which is then thrown over a net ordinarily used to catch the emus. The goal is to land the bone in a pit or hole dug on the other side. Edor and Kee'an, as team-based games that involve throwing, catching, and evasion, can be classified as hunting simulations, helpful for training purposes.

But I stumbled on one strategy game that stands apart. Many believe that it was being played before the arrival of Europeans to the region and is unique to the Maori people of New Zealand.

38: Mu Torere

MU TORERE IS PLAYED ON AN eight-pointed star. The board might be drawn in the sand or carved into wood or bark or marked out with a piece of charcoal. Each player has four counters that are placed on the points of the star. Traditionally counters might be pebbles, bits of vegetables, or, after contact with European travelers, broken bits of china. All that really mattered was that it be clear which counters belonged to which player. This game was simple to make and easy to set up, like many African games we will explore later in our journey.

The players are allowed to move their counters either into the center of the star, called the *putahi*, or to an adjacent empty star point, called a *kawai*. *Putahi* has the association of streams meeting at a junction, while *kawai* refers to a tree branch or even the tentacles of an octopus. The game is won if you block your opponent from moving. They are then declared *piro*, meaning "out" or "defeated," although it can also mean "putrid" or "foul smelling." Interestingly, in modern games we often say a player stinks if he or she plays badly. In order to ensure that the first player can't easily force a win, an additional rule demands that the first two moves by each player be made by stones marked with an asterisk.

The mathematical analysis of how to play this game was first investigated by Marcia Ascher, a pioneer in ethnomathematics who worked at Ithaca College in New York State.

The first question is whether the game actually can be won. Though the game's complexity can make it difficult to know where to start, one clever move to answer this question is to simplify the setup. I often use this strategy to try to crack a puzzle. If we start with just a two-pointed star, rather than an eight-pointed star, and each player has one counter, then the game never stops because there is no way to block your opponent. Either the center or the adjacent star is free to move to. The same happens with a four-pointed star.

Once we start playing with a six-pointed star, things begin to get interesting. Now it is possible for either player to block their opponent. There are thirty different configurations that the board might assume described in the following table. The first entry tells you whose go it is, the second entry tells you whose stone is in the center or whether the center is unoccupied, and the last six entries indicate the arrangement of stones around the star. Configuration (1) or (25) is the starting position for the game according to whether A or B plays first.

(1) AOAAABBB
(2) BAOBBBAA
(3) BAOABBBA
(4) AAOBBAAB
(5) BOBBAABA
(6) ABOABBAA
(7) BBOAAABB
(8) BBOAABBA
(9) AOAABBAB
(10) BAOABBAB
(11) AAOBABBA
(12) BAOBBABA
(13) AAOBABAB
(14) BOABABAB
(15) ABOABABA
(16) BBOAABAB
(17) ABOBAABA
(18) BBOABAAB
(19) ABOAABAB
(20) BBOABABA
(21) AOBABABA
(22) BAOBABAB
(23) BAOBAABB
(24) AAOBBBAA
(25) BOBBBAAA
(26) ABOBAAAB
(27) ABOAAABB
(28) AAOBBABA
(29) AAOABBBA
(30) BBOBAAAB

One can create a diagram that maps out the trajectory of a game on the six-pointed star. The dotted lines indicate moves that can't be made in the opening two rounds.

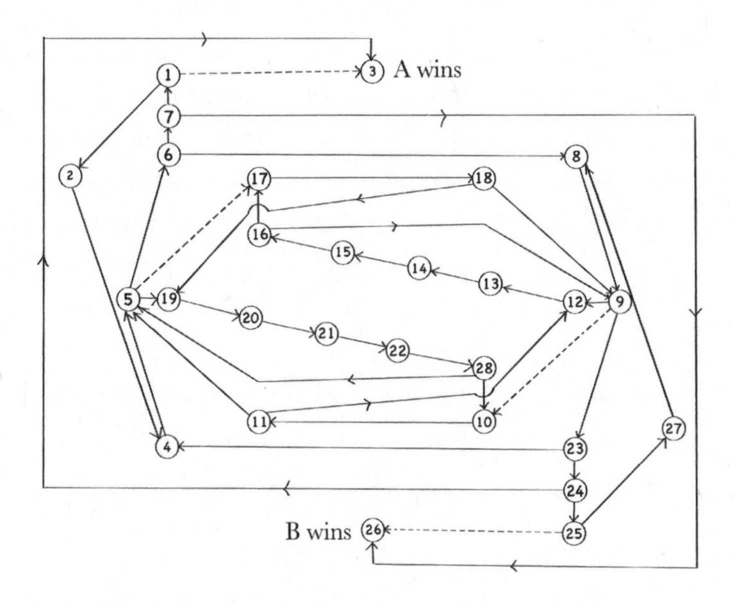

Notice that it turns out you can't get to positions (29) or (30) from the starting positions. And there are only two winning positions, (3) and (26). At certain points you get choices. The junctions in the diagram are key to ending at a winning destination or being forced into a losing configuration. But once again, if players don't make the mistake of choosing the wrong path at positions (7) for B and (24) for A, then the game continues indefinitely.

Once we move on to the eight-pointed star of Mu Torere, the complexity becomes significant. There are ninety-six possible configurations with three different winning positions for each player. As with the six-pointed star, perfect play can ensure that the game goes on indefinitely, but if an experienced player takes on a novice, then the game is sufficiently complex that before too long the inexperienced player will choose the wrong path through the network of moves, and the experienced player will be able to head toward the winning configuration. If you add more arms to the star, the game actually becomes simpler again because the first player can always win on their fourth move. For the game to become fun again, a new restriction would have to be added to avoid an inevitable early victory.

An account of the game written in 1856 by English physician and colonial administrator Edward Shortland describes the way crowds would gather to watch players moving their stones around the board with such intensity and focus. He reports how Europeans were never able to beat Maori players, however hard they tried. Though settlers may have replaced most of the indigenous games with European ones, when it came to games like Mu Torere, it was they who couldn't compete.

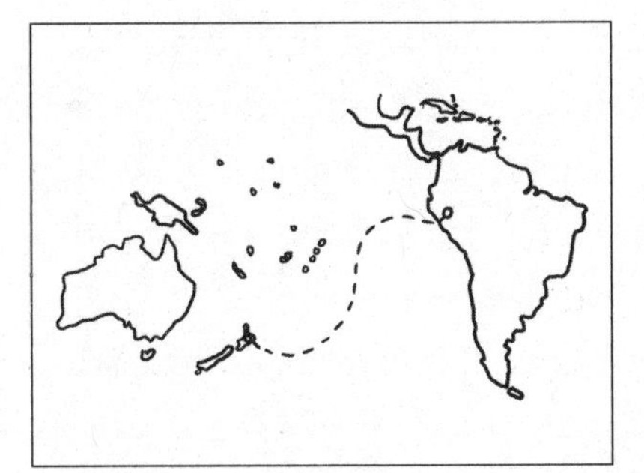

CHAPTER 10

THE SOUTH PACIFIC OCEAN

Games in the Arts

FROM NEW ZEALAND OUR journey will take us next to South America, where we will encounter another story of how indigenous games have battled to survive the influx of those brought from abroad, this time by the Spanish conquistadores. Several of the games we will encounter on our visit to South America demonstrate the importance of dramatic performance by players as part of the competition, the game verging almost on theater. So on this second part of the passage across the Pacific Ocean, I want to explore the interesting dialogue that games have not with mathematics but with the creative arts. For me, games tell stories. But sometimes stories create games.

39: Cluedo

COLONEL MUSTARD IN THE library with the lead pipe. Either this will mean nothing to you, or it will bring a rush of nostalgia for a game that captures a very British fixation with murder mysteries.

Cluedo (or Clue, as it became known in the United States) was inspired by the obsession of its creator, the wonderfully named Anthony

Pratt, with Agatha Christie novels and a popular parlor game called Murder. Pratt was trained as a musician but would while away time between concerts cooped up in country hotels across England by acting out murder mysteries that the other guests would have to solve. This eventually led to the creation of the board game Cluedo.

A murder has been committed at Tudor Close. Dr. Black's body has been discovered at the bottom of the stairs to the cellar. There are six suspects: Colonel Mustard, Miss Scarlet, Mrs. White, Reverend Green, Mrs. Peacock, and, of course, the character I always liked to choose, Professor Plum, who looked very much like an English Einstein. The choice of colors for the names matches the colors of the wine-bottle-shaped counters that were used. From a strategic point of view, it turns out that I should have chosen Mrs. Peacock as she starts in a position on the board that is closest to most rooms.

The game board depicts the layout of Tudor House with its nine rooms, from the ballroom to the billiard room. And the box for the game includes tiny pewter versions of the potential murder weapons, ranging from a dagger to a candlestick, which get moved around during the game play.

Pratt's game could very well have taken as its model Agatha Christie's first whodunnit story, *The Mysterious Affair at Styles*. Set in a country house in Essex, the story features six respectable upper-class suspects who witness the poisoning of wealthy widow Emily Inglethorp. The murder is solved by Christie's master sleuth, Hercule Poirot. The story even includes a map depicting the layout of the country house, which bears a remarkable resemblance to the Cluedo board.

At the beginning of the game one card from each of the categories (suspect, weapon, room) is hidden away in the murder envelope, while the rest of the cards are distributed to the players. The aim of the game is to use the cards in your hand, plus information that you try to pick up during the course of the game about cards in the other players' hands, to deduce the three hidden cards.

You might gain new information by spreading a rumor that the culprit might be Mrs. White in the kitchen with the rope. However, to make the accusation, your counter must be in the kitchen, which is why planning your route around the map of Tudor Close also becomes an important component of good game play. The board adds an element of hindrance to the speed with which you can make accusations about the location of the murder, and it is in fact possible to play the game without using the board at all. Once an accusation is made, if the player on your left has any cards that might disprove the rumor, they must show one of them to you. If they can't show you anything, then the challenge passes to the next person along.

Even if you weren't doing the asking or showing, you could still deduce things from the exchanges that occurred. For example, if I held the kitchen and rope cards, then the other player must have shown the accuser the card depicting Mrs. White.

Given that there are nine rooms, six suspects, and six murder weapons, 9 × 6 × 6 = 324 different scenarios are possible. At one stage in my youth, I got so obsessed with playing this game that I started keeping a tally of the scenarios we'd encountered in the hope that one day we'd have done them all. Life intervened before I got very far through the list of 324 scenarios.

I think I used to love this game so much as a kid because it totally appealed to my burgeoning mathematical passions. The joy of logically eliminating suspects and murder weapons based on the mounting evidence was my first experience of constructing a mathematical proof. I hated it if my sister made a random guess at the solution without having pieced together all the proof for which three cards were missing. But actually this was often clever game play, especially when she could see that I had deduced the answer and was on the way to the location to test my solution. Better to jump in first and make a guess based on her remaining options.

In order to get the most information from your guess, it is important to make sure you don't end up seeing a card that your neighbor has already shown you. For example, you might name a suspect and weapon card that you have in your hand to force your opponents to show you a room card. Games like Cluedo are sometimes called knowledge games because, instead of money, as in Monopoly, or territory, as in Go, players are collecting knowledge as they play.

I used to love trying to lead people astray by spreading a rumor about cards that were actually in my hand. But it turns out that you need to choose carefully which category you are using to bluff. In 2019 two computer scientists, father and son team David and Kyle Hanson, ran twenty thousand computer simulations where one player uses a strategy of bluffing against a collection of honest AI players. Since there are nine room cards and you can only ask about a room if your counter is in that room, these cards are much harder to pin down. So if you spend your time bluffing using the room cards in your hand, you limit how much you can learn. The simulations revealed that doing so decreases your chances of winning by 1.3 percent. In contrast, bluffing on the weapon or suspect cards does in fact give you an edge by increasing your chances by 1.4 percent.

The nature of the game was so original that Pratt was able to successfully secure a patent on it. This was notable, because while most of the time a game inventor might secure a copyright, getting a patent requires proving that nothing similar was created in the past. Pratt's original name for the game was simply Murder! But when the game manufacturer Waddingtons promoted the game shortly after the end of World War II, they decided to market it with something a little less macabre and renamed the game using a combination of Clue and Ludo. The murder was downgraded to "an act," and the original idea of printing a bloody fingerprint on the box was dropped. Due to rationing on cardboard following the war, it took some time before the game hit the high street, but once it did, it very quickly became one of the most popular games of the twentieth century.

Although Pratt very sensibly patented the game, the story is one of those classic cases where the inventor cashes in too early to reap the benefits of his creation's success. Pratt sold his stake for £5,000 only for the game to become an international runaway success, selling 150 million copies across forty countries. Pratt might be good at creating board games, but he was naive when it came to playing the economic game of life. But he wasn't bitter: "A great deal of fun went into to it. So why grumble?"

The murder mystery has proven to be an enduring inspiration for a whole range of modern games, though many dispense with player-versus-player competition and challenge the whole group to solve the challenge together. Sherlock Holmes is the muse for a range of games like The Thames Murders, where players work together to solve a murder mystery. EXIT: The Game has similarly put together a range of collective challenges for players based on the popularity of escape rooms. Solving challenges is rewarded with more clues about the murder. In what would be considered sacrilege for most replayable games, this one actually encourages players to cut up or draw on the cards—after all, once you solve the challenge, there's no point in playing the game again.

A Cluedo Puzzle

Alice and Bob are playing Cluedo, and both have worked out the room that the murder was committed in. There are five possible scenarios left for the weapon and suspect:

1. Colonel Mustard with the candlestick
2. Miss Scarlet with the candlestick
3. Miss Scarlet with the rope
4. Mrs. Peacock with the rope
5. Mrs. Peacock with the revolver

Charles, who has already looked at the cards, tells Alice the suspect, and Bob is told the weapon. Charles is then intrigued. "Can any of you work out the solution?" Both Alice and Bob say no. Then Charles asks again, "Now can any of you work out the solution?" Again both say no. But on the third time he asks, both shout out the correct answer. What do they shout? (I'll give you the answer when we arrive at our destination.)

It's not clear if Agatha Christie ever played Cluedo, but she certainly was fond of games. Her particular poison was bridge. She used it not just as a way to wind down after a day's writing but also as a clue to whodunnit in her stories. Nearly a dozen of them involve a game of bridge at some point in the drama, and generally the game has something to do with the crime.

Take *The King of Clubs*, one of the many stories involving her great Belgian sleuth Hercule Poirot. A game of bridge is meant to provide a great alibi for the four members of the family at the center of the story. But when one player casually mentions the bid made based on the hand of cards dealt before they were rudely interrupted by the shout of "Murder!" Poirot is rather perplexed. Looking at the cards left on the table, the bid doesn't seem to make sense: "She should have gone three spades."

The erroneous bid gets Poirot thinking. As he collects the cards to put them away, he discovers the king of clubs is still in the box. This means that the game was being played with fifty-one cards rather than fifty-two. But fifty-one cards would not be evenly dealt between four players. There is no way they could have played hours of bridge and not have noticed the missing card. The game was never played. Sure enough, the cards were laid out as a decoy and an alibi for the family to commit the murder.

In her novel *Cards on the Table* the game of bridge is again central to the story. This time the game is actually played, and it is one of the four players who has committed the murder. In this story Christie exploits a curious feature of bridge, where one person always lays down their hand for all the players to see, and their cards are played by their partner.

The player who isn't playing is called the dummy. But since they aren't playing, this gives them the chance to slip away from the game, commit a murder, and return in time for the next round of the game.

The clue to who slips away is hidden in the score cards for the game, which keeps track of the bets made and the games won. Christie even published the score cards in the book so that readers could examine them like Poirot and try to deduce the murderer's identity. The key turns out to be one hand in which a player forces their team to make an outlandish bid to win all the tricks, called a grand slam, which the partner has to execute. Such a game becomes a riveting one to play as every trick must be won. The concentration involved in such a feat means that the three active players completely miss that the dummy player has committed the murder.

40: Azad and *The Player of Games*

THERE IS A GAME THAT I have wanted to play ever since I read about it, but I know that I never will—because it doesn't actually exist. Called Azad, it is the creation of novelist Iain M. Banks, described in his science fiction novel *The Player of Games*.

It is fairly common for fantasy novels to cook up fictional games for their protagonists to play. Exploding Snap and Wizard's Chess in Harry Potter and the strategy game Cyvasse in the Game of Thrones are among the myriad examples. But Banks's game is more than just a distraction for the book's characters. The game is used by the Empire of Azad to determine the political, social, and economic structure of the whole planet, in the same way that a democracy has the chance to change direction after every term. The game is played on the turning of every Great Year, which happens every sixth year. Initially twelve thousand players start the game, but this number is gradually whittled down from one round to the next. The game is considered a model of life itself. "Whoever succeeds at the game succeeds in life." Indeed the prize for winning the game is to become emperor.

Along the lines of political manifestos proposed in an election, the game allows players to try out different philosophical and economic models. Once successfully implemented in the game, they can then be transferred to the reality of ruling the empire itself.

The game is played on three huge boards, which are more like landscapes than regular playing boards. They are known as the Board of

Origin, the Board of Form, and finally the Board of Becoming. They stretch out across the halls where the game is played, with parts of the boards rising like hills and dipping down into valleys. The floor is not a regular patchwork of squares but rather a "stunningly complicated and seemingly chaotically abstract and irregular mosaic pattern."

The hills in the board are actually a series of smaller boards stacked like a pyramid in a bewildering meta-pattern. In addition to the main boards there are numerous minor boards where a range of minor games of cards and dice are played out. The pieces are genetically engineered constructs. Part vegetable, part animal, they have the ability to change character as they are moved around the board.

As well as determining the political direction of the empire, the game is also used for gambling by the residents, but the range of bets verges on the barbaric. As a drone explains to the central character of the story, "One might bet, say, the loss of a finger against aggravated male-to-apex rectal rape." In addition to males and females of the species, there is a third intermediate sex called *apex*, which is the dominant sex of all three on the planet.

Apparently the introduction of such barbaric betting was regarded as a liberal move to allow those without money to challenge the rich ruling classes on a level playing field.

I guess that this book always appealed to me because of its conceit that a society's games are a reflection of its culture, ethos, and philosophy. Perhaps one might regard all the games that we have created and play on this planet collectively as Earth's version of Azad. After all, don't we, too, use our games as a way of exploring different models of social interaction? Weren't great chess games between the East and West a way of playing out the Cold War? Isn't the rise of role-playing games or collaborative games, where players work together to beat the game, a reflection of changes in society? And in the games of Go played between man and AI, aren't we getting a glimpse of the role that artificial intelligence is likely to play in the future of humanity?

Perhaps it's not so crazy to wonder if this game might one day be brought to life, especially given that one of its other fans has a reputation for making the impossible happen. Elon Musk so loves the game that he named two of SpaceX's autonomous spaceport drone ships after two ships in the book: *Just Read the Instructions* and *Of Course I Still Love You*. Other options that Musk could have gone for are *Kiss My Ass* and *So Much for Subtlety*. Keep your eye out for those.

41: Games and Riddles

NOT ALL GAMES IN novels are fictional. As we've seen, *The Master of Go* by Yasunari Kawabata uses the famous game of Go between renowned master Shusai and the newcomer Otaké to critique the clash of old and new cultures in Japan. *A Chess Story*, Stefan Zweig's last book before his suicide in 1942, uses a game of chess aboard a ship from New York to Buenos Aires to explore Nazism and the psychological effect of persecution and obsession. Several centuries ago, riddles often played a part in storytelling, and at times what we would recognize as games would even serve a critical role in a narrative.

In the Hervarar Saga, a thirteenth-century legend about the wars between the Goths and the Huns, the disguised god Odin sets King Heidrek a sequence of riddles.

> Who are the maids that fight weaponless around the lord, the brown ever sheltering and the fair ever attacking him?

The king is a keen gamer and easily answers. Odin is talking about the pieces in the Norse game of Hnefatafl. The 13 × 13 board has an odd number of squares, allowing for a central position, which is occupied by the king. It is surrounded by guards, or as Odin refers to them, "maids that fight weaponless." Positioned on the outskirts of the board is the opposing army. We glean from Odin's question that, like in chess, each army had its own color: fair and brown.

What is rather unique about this Norse game is that the opposing army is twice the size of the defending army. In the next chapter we will encounter a South American game called Adugo that has a similar asymmetry in pieces: a single jaguar is pitched against many dogs. In the Norse version, the aim of the king is to escape to the edge of the board while the opposing army tries to trap the king's movement.

Odin follows up with another riddle related to the game:

> What is the beast all girdled with iron which kills the flocks? It has eight horns but no head?

The *hnefl*, replies King Heidrek, referring to the king piece sitting in the middle of the board surrounded by eight squares on which the guards sit.

This saga is thought to have influenced J. R. R. Tolkien in writing *The Lord of the Rings* and *The Hobbit*. Riddles were an important part of Bilbo Baggins's journey when he encountered Gollum, and those in the story could well have been inspired by the thirty-seven riddles that Odin set King Heidrek.

Hnefatafl pieces have been discovered as far afield as the Orkney Islands, Iceland, and Ireland, and across Scandinavia. That they were often found in burial chambers implies that the game was more than just a fun pastime to while away a dark evening but, like the Egyptian game of Senet, had religious significance for the journey to the other side. They indicated the status of the deceased as a powerful warrior, skilled in strategy and therefore to be respected.

Although many of these pieces are simply made of carved ivory, a rather fine gaming piece emerged in September 2019. A group of archeologists were digging on Lindesfarne, an island off the Northumberland coast, famous for the early medieval monastery that was raided by the Vikings in 793 AD. It was home to some of the most beautifully illuminated gospels created in the early eighth century.

But dating from the same period as the gospels, the archeologists were very excited when among the pots and bones, the mother of a team member, visiting to help celebrate her daughter's birthday, dug up a beautiful blue-glass gaming piece with white swirls and five white beads arranged like a crown.

This is believed to be the king piece from a game of Hnefatafl. It's not clear whether the piece was already on the island before the Vikings invaded or was brought by a game-loving raider. Perhaps the monks weren't just dedicating themselves to the austere life of the church but were enjoying a few games of an evening. To get a glimpse of the games played on this holy island is truly exciting. The director of the dig summed up the thrill felt by many on discovering this gem: "My heart was pounding, the little hairs on my arms were standing up. As a scientist, you have to train yourself out of having an emotional response to things like this. It's a piece of evidence, bottom line. But honestly. It's just so beautiful and so evocative of that time period, I couldn't help myself."

42: Theater Games

NOT ONLY HAVE GAMES played a part in works of literature, but they have also been used as crucial tools in helping to stimulate ideas across many of the creative arts. I think one of the reasons I was drawn to Dungeons & Dragons all those years ago was the role that performance played in the game. It was a game that appealed to the thespian in me. I was not alone, as it turned out.

Besides math and games, I have always had a great passion for theater. When I was younger, I thought that theater was the rather simple combination of an author's text and a group of actors' performance. But it wasn't until I encountered the theater company Complicité for the first time at my local community theater in Oxford that I understood the role of play itself in making theater.

If you are making a new piece of theater or improvising a scene, then the bare stage, empty set, or blank page can be terrifying and completely kill off the chance of creative theater making. Inexperienced actors become very self-conscious and awkward if challenged to make something up in front of others. Rather than total freedom, what often helps creativity is to put on constraints, rules that push someone in an unexpected direction. This is when a good theater game can be really helpful in providing a framework to start playing around.

For a start, a game is often totally absorbing, and the simple set of rules often gives you something to fall back on if ideas are not forthcoming. It encourages people to work as a team, collaborating, but also involves a bit of competitiveness, which can help create a sense of energy that can be exploited. Winning or losing starts to elicit natural, spontaneous emotional reactions from the actors. We experience drama. We don't know what is going to happen. The rules might be written, but the way the game is played should change each time.

In the theatrical pieces that I've created together with actors, we always begin each day with a game of nine square. It's an enlarged version of the game played in schoolyards called four square. In our expanded version we mark on the floor a 3 × 3 grid and number the squares in a spiral with nine in the middle. The king of the game is the person standing on the central nine square. They serve the ball by bouncing it into one of the other squares. These are defended by the other actors. If the ball bounces in your square, you must pat it into someone else's square. If you miss the ball or knock it out of the court, then you get sent down to square one, the beginning square, and everyone behind you moves one square up. Your aim is to try to get to the nine square and be king. The game is very addictive. When making theater one needs to be careful not to use up the rehearsal time playing this game. But the time spent playing is worth it because the game acts as a very effective way of getting actors to pay careful attention to the actions of their fellow actors. The power of games is well appreciated by theatrical actors and teachers. Many of those at Complicité trained at the École internationale de théâtre Jacques Lecoq, a school of physical theater founded by Jacques Lecoq in 1956. Lecoq talked often of the importance of *le jeu*, "the game." All games involve reacting to the moves of your fellow players, offering a new move in response that contributes to the dynamic arc of the game being played. For Lecoq *le jeu* created an environment where the actor could not perform alone, letting their ego get in the way, but had to respond to the play around them. A game also encourages being present in the moment of play. It creates joy in the actor.

One example I often use serves to show the power of mathematics as a tool for generating interesting dynamics in a group of actors. In a group of ideally about twenty actors, you begin by asking each actor to pick two other actors in the space while not actually letting them know they have been picked. When I say "go," each actor is to move so as to create an equilateral triangle with the two other actors they've chosen. This means moving so that the distance between you and the two actors is equal to the distance between the two actors.

The trouble is that as soon as you say go, the other two actors you've chosen start moving to create equilateral triangles with the actors they've chosen. It is a very simple algorithm, but the result is chaos. From the outside as an audience member, it would probably be impossible to decipher the rule underlying this wild crowd. The rhythm of the movement is fascinating because it pulses between fast-paced moments as the triangles change quickly and then moments of calm as things seem to stabilize. But like a mathematical chaotic system, a small butterfly flapping its wings can have big impacts. As one person adjusts to make their triangle a more perfect equilateral, another shifts to take account of the small change, and the small movements swiftly cascade into another period of wild motion.

It would be almost impossible to choreograph such a scene in a top-down manner, the director giving each actor individual instructions. Instead the bottom-up nature of the game means a simple rule can result in complex outcomes. For Lecoq, *le jeu* involved "an improvisation for spectators using rhythm, tempo, space, form." Mathematics is a brilliant way to create games of rhythm, exploring the nature of space; it is the subject of form and is the language for understanding a world in flux. For Lecoq his work is "an attempt to discover abstract, essential principles underlying all lived experience." That sounds not too dissimilar to what I do as a mathematician.

Lecoq comments in his book *The Moving Body*, "I am sometimes told 'Because it is so structured, we have no freedom.' The opposite is true, however." Within the rules of the game, one is pushed out of the comfort zone of one's daily thinking and into somewhere new. The best games we've invented have that quality: the constraints of the rules still provide the platform for great creativity.

43: Mozart's Dice Game

ANOTHER KIND OF NARRATIVE without words that also involves mathematics is music—and music can be created and enlivened by games just as much as theater can. I have already hinted at how the *I Ching* was used by John Cage to determine how to play one of his pieces. But he was not the first composer to use the idea of a game of chance to write music.

In his *Musikalisches Würfelspiel*, or "musical dice game," Mozart proposed a cunning way of generating a sixteen-bar waltz using a set of dice. First published in 1792, a year after Mozart's death, the piece consists of 176 bars arranged in a 11 × 16 configuration. By throwing two dice and subtracting one from the score, you run through the sixteen columns picking out which of the eleven bars will be played. The mastery of the composition is of course to produce 176 bars that fit perfectly together into a convincing waltz whatever the throw of the dice. The mathematics of probability indicates that some rolls of the dice are much more likely than others: for example, there is a one-in-six chance of rolling a seven, as opposed to a one-in-thirty-six chance of rolling a twelve. So some bars of music in Mozart's game are more likely to be played than others.

The staggering thing is that there are 11^{16} different waltzes that you can generate using this system. That's forty-six million billion waltzes. If they were played one after the other, it would take two hundred million years to hear every one. If I were to get my dice out and start rolling a waltz, you would likely hear a Mozart waltz that has never been heard before in the history of the planet. A world premiere!

Before we disembark to explore the games of South and Central America, I must return to Cluedo and deliver the solution to my puzzle: they both shout out, "Miss Scarlet with the rope."

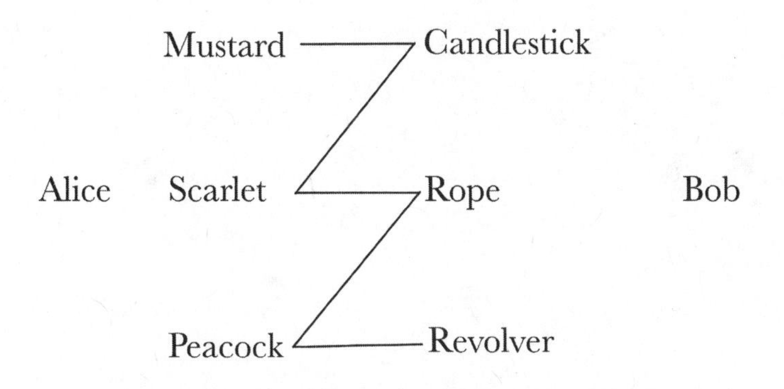

How do they work it out? Here is a picture whose lines show the five possibilities. Alice and Bob have to work out which line represents the solution. Charles, who has already looked at the cards, tells Alice the suspect, and Bob is told the weapon. Charles then asks them for the first time, "Can any of you work out the solution?" If either had been shown a weapon or person with a single line emanating from them, then that person could have worked out the solution. This was not the case. So they can both delete the top and bottom line. In this new diagram with three lines drawn, again, if either Alice or Bob is shown a weapon or person with a single line, then on the second time Charles asks them, they could deduce the solution. But neither shouts out. So again they can both delete the single lines in this revised graph. That just leaves the one remaining line connecting Miss Scarlet to the rope, which they both shout out on the third time of asking.

As our journey lands on South American shores, we will remain focused on morbid subjects, moving from murder mysteries to the Day of the Dead.

CHAPTER 11

SOUTH AND CENTRAL AMERICA

MY TWO DAUGHTERS WERE adopted from Guatemala, so it has been especially important for me to seek out the games of this region to celebrate their heritage and culture while growing up in London. They are both exuberant, passionate girls, which I think describes well the games and festivities we encountered during the time our family lived in Central America. There is one holiday in particular in the calendar year that we all enjoy celebrating with dressing up, eating, and, of course, playing a game.

44: Mexican Bingo

DÍA DE LOS MUERTOS. The Day of the Dead. What an extraordinary celebration: parties, picnics, drinking, music, and dancing on the graves of your ancestors.

I was visiting Mexico as part of work I was doing with Mexican artist Eduardo Terrazas, designer of the iconic logo for the Olympics in Mexico in 1968. Terrazas and I have been collaborating on an amazing design that Terrazas has been working with for the last four decades, which he calls "The Cosmos." His simple diagram captures the physical fabric of the universe: from the four fundamental forces to the geometry of space. But Terrazas was very happy to combine our work with a trip to Oaxaca to celebrate the Day of the Dead.

Attending the fiesta in Mexico had been on my bucket list for my journey around the world for years, but only recently did I have the chance to fulfill that dream. The depiction of this festival in the wonderful animated movie *Coco* had only whetted my appetite further. Judging by the many Japanese faces in the graveyards of Oaxaca the year I went, the film *Coco* must have had a similarly captivating effect on cinemagoers in Tokyo.

Before heading out to party, Terrazas and I dropped in on a local makeup artist who painted our faces with the traditional *calavera*, or "skull," imagery associated with the festival. After an hour of careful painting, we were ready to head to the graveyards to join the festivities.

The Day of the Dead is no somber affair but a joyful celebration of the lives of those who have passed on. Graves are adorned with marigold flowers, *pan de muerto*, sugar skulls known as *calavera de azucar*, and beverages that the dead once liked. We came across one grave that featured the most extraordinary flower sculpture of a man playing a piano, celebrating the musicianship of the family's ancestor. We followed mariachi bands around the local villages, pumping out music until a glow began to appear in the sky to the east, and we collapsed in our beds.

I was keen to pick up some Day of the Dead memorabilia to take home, and the next day I stumbled upon the game of Mexican bingo, or Loteria. Terrazas told me that every Mexican household has a set of Loteria cards, as it is one of the traditional games played across the country. Loteria consists of fifty-four cards with images from Mexican culture, which the nominated cantor, or caller, will deal out in turn, announcing the name of the card as it is drawn. There is an interesting amount of creativity, even poetic imagery, involved in the cards' names—"the coat for the poor," for example, is the description of the sun card.

Each player is given a *tabla*, or "board," with a 4 × 4 grid of pictures that they will try to match with the cards being called by the cantor. Complete a row or column, and you can call *loteria*. My set of cards was full of the imagery from the Day of the Dead. The characters reminded me of many of the figures that appear in the Major Arcana in the tarot. *La luna*, the moon, "the street lamp of lovers." *La muerte*, death, "thin and lanky." But other sets of Loteria cards include characters that range across the natural world and the domestic domain.

This quintessentially Mexican game actually has its origins in Europe. The classic set of Loteria cards dates from their introduction from Europe to Mexico in 1769, but more modern sets have updated the characters to celebrate Mexican culture. The cards that have become standard were first published in 1887 and played an important part in representing and normalizing different aspects of Mexico's national identity during the nineteenth century. They were even distributed to soldiers alongside the rations they received. Since then, new cards have found their way into the deck. My set, for example, includes cards depicting the Mexican artists Frida Kahlo and Diego Rivera. Since I brought the game back home, it has become a family favorite, far more engaging than the British version of bingo.

The version of bingo that I grew up with also arrived on British shores from Europe in the eighteenth century, but bingo halls in the United Kingdom really only came into their own in the 1960s, when the Betting and Gaming Act allowed large cash prizes to be made available for completed cards. Dance halls and cinemas converted overnight into bingo halls to accommodate the growing demand for seats at the bingo tables.

Punters would buy cards arranged into three rows and nine columns with fifteen numbers arranged across the grid. They could buy a set of six cards with the guarantee that every number appeared somewhere on the cards. How many balls would you expect to have to draw before someone says bingo? If there are fifteen hundred cards in play during a game, then the mathematics says that you would expect as many as fifty-six balls would need to be called before a punter shouts bingo. A similar calculation for my Loteria de Muerte revealed that, with five of us playing, we would expect twenty-one out of the fifty-four cards would need to be drawn before we were likely to see one of our rows or columns completed.

5				49		63	75	80
		28	34		52	66	77	82
6	11				59	69		

The Math of Bingo

In most bingo halls it doesn't matter how long it takes you to match all your fifteen numbers; the payout is the same. But shouldn't you be rewarded if you are lucky enough to get all fifteen numbers in a shorter-than-expected time? In some halls there is a bonus prize if you shout bingo in forty numbers or fewer. If you are devising a sliding scale of prize money, then mathematics will be very useful in making sure the price of a card balances the expected payout.

Suppose a bingo hall pays £100 for the winner and charges twenty pence per card. Fifteen hundred cards per game makes £200. But what if you changed the game? For example, if you paid out only £50 for bingo called in more than fifty numbers, but increased the prize for covering all your cards in fewer numbers—bingo in forty-nine wins £100, bingo in forty-eight wins £200,..., bingo in forty-one wins £900, and bingo in forty or less wins £1,000—then the expected payout in a game comes out at £63 if fifteen hundred cards are being played. At twenty pence a card, the bingo hall is making more money. But punters would probably pay more knowing they might have a chance of winning £1,000 if they are lucky.

In the United States, bingo cards consist of 5 × 5 grids (with the central square not filled) with numbers running from one to seventy-five. The first column is restricted to numbers from one to fifteen. The second from sixteen to thirty, and so on, till the final column has numbers from sixty-one to seventy-five. Since the first column chooses five numbers from fifteen, there are $\frac{15\times14\times13\times12\times11}{5\times4\times3\times2\times1}$ different ways to choose the first column. The same is true for the other columns, except the middle one, where you choose four numbers from fifteen, giving $\frac{15\times14\times13\times12}{4\times3\times2\times1}$ different columns. This make a total of 111,007,923,832,370,600 different lottery cards—more than enough to go around.

Popular as the game became in countries like the United Kingdom and the United States, the European version lacked the artistry and poetry of Mexican bingo. Callers would read numbers from one to ninety off balls—a far cry from intoning words like *el corazón* and *la luna*.

45: Jogo do Bicho

WHILE A CALL OF "THE DEER" might elicit a response of bingo around the family dining table in Mexico, down in Brazil the animal has more serious implications. Jogo do Bicho, or "the animal game," is big business in Brazil.

The game is simple. There are twenty-five animals, and players bet on which is going to be drawn from the selection. Simple enough—yet despite its harmless description, gambling on the game has become such a problem that for many years playing the game has been banned in twenty-five out of the twenty-six states in Brazil.

It all started off so innocently. In the late nineteenth century, the Baron de Drummond had a clever idea for drumming up publicity for his zoo in Rio de Janeiro. On each ticket that was sold to visit the zoo, there would be a different animal printed. At the end of the day Drummond would reveal an animal that he'd hid behind a curtain, and anyone whose ticket matched the chosen animal would win a prize. The game proved hugely popular and began to spread beyond the zoo.

Entrepreneurial types realized you didn't need the actual animals behind a curtain and started to give punters the chance to bet on a list of twenty-five animals. Known as *bicheiros*, they would walk around the city with placards bedecked with images of the twenty-five animals, selling tickets for that day's draw. Each animal had a set of four numbers attached to it. The ostrich had numbers one, two, three, and four. The deer had ninety-three through ninety-six, and the cow ninety-seven through one hundred. The game became a mix of animals and numbers. You could bet on an animal with odds of twenty-five to one, but if you were feeling more daring, you could bet on the number of the animal at odds of one hundred to one.

One appeal of the game was its unregulated character. Unlike with the state lottery, you were free to bet—and lose—as much as you wanted. As animal fever began to grip the country, the authorities decided to clamp down on this explosion of illegal gambling. They'd been OK with Drummond using the idea to raise money for the zoo, but now huge sums were passing through hands on the street—and the government wasn't seeing any of it. Ever since that first attempt in 1895 to ban the game, the authorities in Brazil have been wrestling to stop the gambling on these twenty-five animals, but with millions still playing the game every day, it has proven almost impossible to contain.

It is believed that half a billion dollars is exchanged each year through the game, which employs roughly 1 percent of the nation's total workforce. The draw for the game often happens in the local offices of the *bicheiros* at 2 p.m., and then the news of the winning animal is quickly broadcast across the city. Phones are in such heavy use communicating the news that telecommunication companies call the hour from 2 to 3 p.m. "the *bicho* hour."

Some have tried to attribute the popularity of the game to connections with the potency of animal imagery in indigenous culture. But others argue that its success is a more modern phenomenon, a product of the exploding urban environment that emerged in Brazil at the end of the nineteenth century. In short, this was a vehicle for capitalism, not shamanism.

Certainly superstition and dream interpretations have played a significant part in the journey of this game. Many will credit a dream of an animal with their decision to bet on that particular beast. Each animal has built up its own associations beyond the animal itself. A dream of a naked woman would mean betting on a horse. If a Portuguese man visits your dreams, then bet on the ass. The elephant became associated with death. If there was a traffic accident involving a car whose registration number had the numbers forty-five to forty-eight, the numbers of the elephant, that week would see huge bets being placed on the elephant. A derailed locomotive featured in the newspapers in the 1960s caused such a run on the numbers appearing on the train that the *bicheiros* were fearful that they wouldn't be able to pay out if the numbers came up. Although outside the law, the *bicheiros* are generally credited with treating the game with great honesty, and trust between the players and *bicheiros* is very high.

Individual *bicheiros* might offer their own odds on particular combinations of numbers and animals occurring in draws. Understanding the odds and setting rewards accordingly meant that barely literate punters were able to do levels of arithmetic and assess probabilities well beyond what the average student in school might expect to be able to master. As a result the game is responsible for a subtle knowledge of the laws of probability and arithmetic permeating the population.

The impact of this game goes well beyond simply the daily gambling. To declare that something is as likely as a zebra being drawn in Jogo do Bicho is the Brazilian way of expressing the idea of a huge upset. The zebra does not occur among the choice of twenty-five animals. The deer, however, does. It is the twenty-fourth of the twenty-five animals on offer. But because the deer is traditionally associated with homosexuality, macho Brazilians have tended to shun the number twenty-four. For example, football players have been known to refuse to play in the twenty-four shirt, even when playing internationally in games that require the use of sequential numbers. Politicians in the Brazilian senate have cars numbered from SF-0001 to SF-0095; however SF-0024 is missing from the fleet. Conversely LGBT groups will exploit the association by including the number twenty-four in their campaign materials.

The profits made by the *bicheiros* are often fed back into the community in a sort of informal economy. A number of football clubs benefit from support from the game. Treze Futebal Clube sports the rooster as its mascot because *treze* means "thirteen" and the rooster is animal thirteen in the list of twenty-five. Even the carnival in Rio profits from the game as the *bicheiros* are the major patrons of Rio's samba schools, the training ground for performers at the festival. This parallel economy is basically run by the mafia in Brazil, and the authorities are having a difficult time dealing with the Brazilian love of the game of animals.

The senate in Brazil is now considering giving in and just legalizing the game. As Senator Angelo Coronel remarked, "Today it is a custom. They play from evangelicals to priests, from doctors to lawyers, from policemen to politicians. A large part of them bets on it, but they do it clandestinely. It's time for us to legalize it."

46: Adugo and Komikan

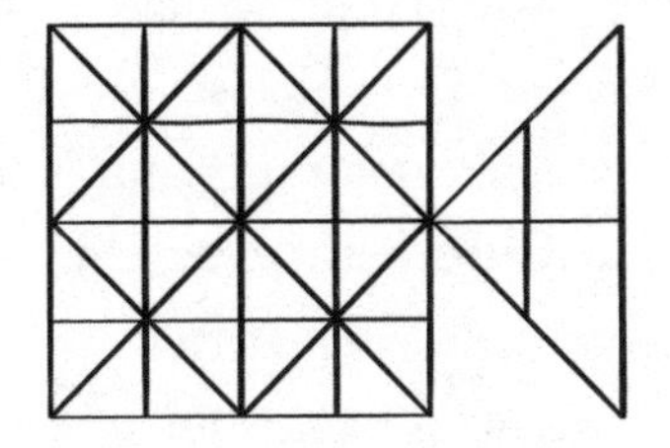

TRAVELING IN SOUTH AMERICA on the hunt for games comes with a set of challenges. As was the case in New Zealand, I am less interested in the European imports that have blotted out older, indigenous games. Seeking games that can't be traced to the arrival of the Spanish conquistadores to this region, I finally found some success when I turned up two similar-looking games called Adugo and Komikan. Their intriguing setup adds a triangle to the edge of the usual square boards that I've come across.

The other rather unique trait of these games is the asymmetry involved. In chess or checkers, or nearly any other game, each player has the same number of pieces. The idea is that players should start on an equal footing. But in Adugo and Komikan, one player has one piece, while the other player has many. To compensate for the inequality, the single piece has a more flexible range of movements compared to the other pieces. Most of the variations on this game are based on the idea of the single piece being a wild animal, a predator, like a jaguar. And the pieces belonging to the other player are dogs hunting the wild animal or domestic animals, like goats, that the predator tries to attack.

The jaguar can capture the dogs by jumping over them, like in checkers. The dogs have more limited movements and can just slide one step along the lines of the board. The challenge is for the jaguar to capture five out of the fourteen dogs before the dogs surround the jaguar, preventing it from making a move. The version of the game called Adugo is generally regarded as having been created by the Brazilian indigenous Bororo people from the Pantanal region.

Adugo is their word for "jaguar." Originally it seems the board was drawn in the dust on the ground, and the pieces were little stones. The Mapuche tribe, indigenous to central Chile and Argentina, refers to its variant of the game as *Komikan*, which means "to eat all."

Probably the best evidence for the game having pre-Hispanic origins is the discovery in Chinchero, near Cuzco in Peru, of a board scratched into a wall that dates from before the Spanish arrived in the region. But there are some who have doubts. The design of the board bears similarities to the board used for Alquerque, a game that originated in the Middle East in the tenth century and was brought to Spain by the Moors some centuries later.

But Alquerque is missing the extra triangular piece that you find in the South American version, and the players of Alquerque have equal pieces, unlike the asymmetric version that captures the hunting of a wild animal. It is unclear if this version grew out of the introduction of Alquerque to South America or evolved independently, but there is no question that Alquerque was a game that the Spanish brought to the peoples they conquered. An illustration dating from 1615 shows the Inca king Atahualpa as he languishes in the Cajamarca jail awaiting his death in 1533. He seems to have been a real game fanatic. Despite a gruesome death waiting just around the corner, he is playing a game on the square Alquerque board with his Spanish guard.

47: Sapo

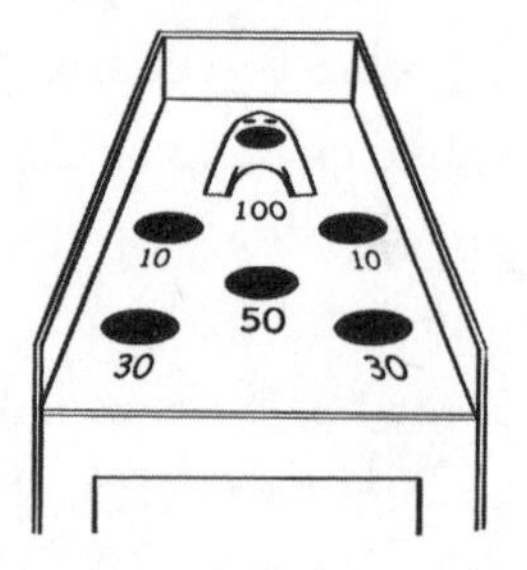

ANYONE TRAVELING IN SOUTH AMERICA has probably come across a Sapo board in a café or bar. The board is unmistakable, with its distinctive array of holes atop a large cabinet. (As much as I wanted to add Sapo to my collection of games collected from around the world, my wife balked at stowing one of these huge wooden cabinets in our hand luggage.)

To play the game, you launch golden coins onto the board to try to get them into the holes. Each hole connects to a different drawer of the cabinet, which scores different points. In the center of the board is a brass toad with an open mouth, which provides the game's name: *el sapo* is Spanish for "toad." Landing the coin in the toad's mouth yields the maximum possible points.

The game is based on an ancient story about the sun god visiting Lake Titicaca in Peru to throw coins to the toad to tap its magical powers. Boards will sometimes feature an image of a large bronze sun as their backdrop as a nod to this story.

Sapo is most probably a game that truly does precede the arrival of the Spanish to South America. Games of throwing are ubiquitous across the world, most often in games designed for training people for hunting and warfare. The variation I grew up with involved throwing marbles in the playground. For the indigenous communities in South America, being good at Sapo would translate into success in the hunt for food—besides keeping children and adults occupied for hours at play.

HARD AS I SEARCHED FOR South American games that predated European contact, I found plenty of interesting games that had taken on a new local flavor once they were introduced by European settlers and traders. Truco is a good example of a game that, although it has its origins in Spain and southern France, is now firmly associated with South America. The game is played with a Spanish deck of forty cards with suits of coins, cups, swords, and batons.

What I love about truco is that the joking and body language that accompany the game are as important as the cards you are dealt. Frankly it often looks like the players are about to start a fight, but the energy is all part of the game.

Played in Brazil, Argentina, Uruguay, and beyond, it is a card game with the aim of taking tricks. It has resonances with bridge or whist because you play in pairs—mostly two pairs but sometimes three. Rather than the whole pack being dealt out, each player only gets three cards. As well as winning tricks, you can score points for having certain combinations of cards in your hand. For example, if you have three cards of the same suit, called a *flor*, or "flower," then you can score points for this provided that another player doesn't have a higher scoring *flor*. Rather than collecting up cards when tricks are won, the players keep their cards in front of them so that combinations they hold in their hands can be scored at the end.

Because banter is an important part of playing the game, people will often use the calls of different combinations as a chance to recite some poetry, not always appropriate for younger ears.

> Pintor que pinto la luna pinto la luna y el sol
> pinto a tu hermana desnuda y en cada teta una flor.

Be careful also what you say in normal conversation during the game. Say the word *flor* by mistake, and you will be held to the claim that all your cards have the same suit. If they don't, you will suffer a penalty. It also takes an idea from backgammon where a team can challenge another team with doubling the points to be won if they think they are in a stronger position.

Because the game is played in pairs, there is the opportunity, like in whist and bridge, to communicate information about your cards to your partner to help judge whether to accept bets or to decide which cards to play. But one of the charming qualities of the South American version is that this communication is done via winks and nods rather than being confined to the play of the cards or calls of bets, as in whist or bridge. And it is quite acceptable to communicate a wink to your partner in such a way that the opposing pair don't see it. Deception is all part of the fun.

The signals vary from one region to another. But it is generally regarded as good form to use a set of signals that are collectively recognized rather than concocting your own secret signals that can't be decoded. For example, in Argentina, to indicate that you've got the ace of swords, you raise your eyebrows. Send your partner a kiss, and that means you've got a two. Bite your lower lip to indicate a three in your hand—important information, because one quirk of the game is that threes and twos are actually considered high ranking when it comes to taking tricks.

At times this kind of private signaling can almost seem like a form of cheating. One imagines that at some point in the game's evolution, it was played without these gestures. Then someone noticed that a couple of players were doing incredibly well, almost as if they were able to read each other's mind. And then the gestures between them were noticed. There were two ways forward: either ban signaling or embrace it as part of the game.

Truco may have given in to signaling, but for games where real money is on the line, tougher lines sometimes need to be drawn. Casinos have had to wrestle with complex questions about when a player is breaking the law or simply exploiting flaws in the game. In March 2004, three eastern European gamblers were arrested leaving the Ritz casino in London after winning £1.2 million. The offense? They'd used an electronic device to implement a mathematical analysis of the roulette wheel to predict where the ball would land. They had had help with their bets, to be sure—but the court case ultimately found them not guilty of cheating because they hadn't tampered with the wheel.

Another example occurred in 2012, when a gambler named Phil Ivey made £7.7 million playing a version of the card game baccarat at Crockfords casino in London. Ivey spotted an anomaly: the edges of the design on the backs of cards are not symmetrical. This is called *edge sorting*, and to exploit it he asked the dealer to turn high cards round when they were dealt with the excuse that he thought it brought him good luck. Once the cards were redealt, Ivey could see from the backs of the cards before they were turned if they were high or low cards. This time, the courts determined that Ivey had cheated, deeming Ivey's actions amounted to interference with the cards.

That is what is so fascinating about cheating. Generally it is not regarded as an acceptable part of playing a game. You've signed up to play by the rules, so what satisfaction is there to winning if you've only done so by breaking them? The reasoning is a bit clearer when money is on the line—cheating can be rationalized by a player because the goal isn't to have fun playing by the rules but to earn as much money as possible. But even when money isn't on the line, some players still cheat. There is a strange tension on display: on the one hand the cheater is breaking the rules by cheating; on the other hand, a cheat is trying to give the impression to opponents that they are playing by the rules, so it's not as if they have abandoned the game's framework completely.

How do you deal with a game where everyone is trying to cheat? In truco, once it was recognized that players were using gestures to communicate information about their cards, what better solution than to bring cheating under the rules of the game. When game manufacturer Hasbro found in research that 50 percent of people admitted to cheating at Monopoly, they felt compelled to embrace the cheating and created a special Monopoly game: the Cheater's Edition. The Community Chest and Chance cards include fifteen special cards that challenge the player to carry out a particular cheat, like steal from the bank or move your opponent's piece to a new square without being caught.

There are some games where cheating is a core part of the game play. Take Cheat!, a card game where you place cards face down sequentially, trying to get rid of your cards.

Someone might declare that they are putting down three kings. Then someone adds to this two kings. Clearly someone is cheating, but the game is to challenge someone when you think they are cheating. Get it wrong, and you pick up all the cards. Catch someone cheating, and they take the cards.

There are certain scenarios where cheating is tolerated, if not incorporated into the game. Finding the cheats hidden inside the code of a video game is for some fair game and part of the fun of hacking a game. It is important to distinguish different kinds of cheats here: there are cheats that are programmed in intentionally, and then there are glitches, hacks, and the kinds of exploits that let players do things the programmer never intended. In the latter case it becomes almost a competition between the player and the programmer. In the former, less so, since the programmer put the cheat in there in the first place—arguably it is an intended feature of the game. There is still the challenge for the player to unearth these hidden Easter eggs.

The use of cheats in some video games is generally an indication that the game has become too difficult, and you are finding the obstacles to advancement decreasingly fair and fun. But by giving yourself infinite lives, you may risk veering in the other direction and making the game too easy, such that the obstacles are no longer challenging enough to maintain your involvement. Video cheats allow you to calibrate accessibility, managing your own level of engagement in the game.

What if everyone abandons the rules, throwing open the doors to cheating of any kind? Such a situation is described in *The Grasshopper*, Bernard Suits's book about games. Two chess players decide that any move should be allowable as they seek to immobilize their opponent's king. This can include actions as outlandish as gluing the king to the board or chopping off your opponent's hands so they can't move the pieces.

Eventually they conclude that what they are playing is a duel to the death. So instead they agree to meet at dawn to fight it out. But is this even still a game? The grasshopper who narrates the story says that there is still one rule they've agreed on: to meet at dawn. This one rule they have agreed to abide by means that they are still playing a game with one rule. If there were truly no rules, one could always get the jump on the other by striking first, well before the appointed hour.

Fortunately, the games I found in South America never developed into a death match in quite this way. But cheating continued to feature heavily in some of the examples that emerged from my journey. One game from Peru is notable for the importance of lying or bluffing to winning.

49: Perudo or Dudo or Liar's Dice

ANCIENT SITES ACROSS SOUTH AMERICA, including Machu Pichu in Peru, have revealed artifacts that hint strongly that dice games were part of indigenous culture. The dice were often counters that landed on one of two sides, for example corn kernels with one side colored black. But there are also examples of dice that look rather like the tall pyramids you see at the ancient Mayan and Inca sites. On each of the four sides of the pyramid would be marked a number of lines to indicate the roll of the dice.

Such dice might have been used alongside a board to keep track of the throws. The archeological site of Tlacuachero in southern Mexico has revealed a mysterious arrangement of holes in a semicircular shape punched into the clay floors. Carbon dating has shown these holes probably date back forty-eight hundred years. Similar arrangements of holes are used by contemporary communities as boards for keeping track of dice throws in games similar to the racing games of ancient Egypt and Babylon. The belief is that the holes were used with games of dice in South America nearly five millennia ago.

One game involving dice that was played by many pre-Columbian cultures is Patolli. The mid-sixteenth-century Aztec Magliabechiano codex shows four indigenous people playing Patolli watched over by the Aztec god of games, Macuilxochitl. The game shares a lot in common with the Indian game of pachisi or Ludo, where pieces race around a cross-shaped board, trying to get home before they are captured by other players. It is a great example of how the same game is invented by two different cultures that never had contact with each other.

Another dice game reputed to have origins in ancient Peru that has become very popular in modern times is known as Dudo, which means "I doubt it." Each player has a cup and five dice that they roll, covering the outcome with the cup. Players then take turns making claims about how many dice of a particular roll there are in total among all the throws. Players can only see their own dice. Seeing three fives in their roll might embolden a player to claim a high number of fives in total. But perhaps they are bluffing and leading others astray. This is why in some circles the game is known as liar's dice. As players take turns, they either up the number of dice they are predicting, switch the call to a different dice roll, or shout Dudo to challenge the last claim made. Once Dudo is called, the dice are revealed. If the caller was right, the person who made the incorrect claim loses a die. If the caller was wrong, then the caller loses a die. If you don't have any dice to hand, you can play another version, called liar's poker, which uses the serial numbers on dollar bills to make wagers.

An old story claims that the Inca king Atahualpa, who, as we saw, loved playing games to the bitter end, showed Dudo to his captor, Francisco Pizarro, in an attempt to gain his freedom. Pizarro so enjoyed the game that he took it back with him to Spain—but not enough to change his mind about Atahualpa's fate: he was garroted, bringing the reign of the Inca kings to a bloody end.

The game received a fresh burst of interest in the late 1980s when it was marketed by Cosmo Fry in the United Kingdom under the name Perudo—a mix of Peru and Dudo. Fry had come across the game while playing golf in Lima—or rather, while not playing golf. Confined to the clubhouse during a rainstorm, the locals brought out the dice and started teaching him the game. He found the game addictive and decided to bring it back to Europe. He wasn't the only one who loved betting on the dice. The game starting flying off the shelves when Stephen Fry (no relation apparently) endorsed the game as "the second most addictive thing to come out of South America."

The game makes a prominent appearance in the movie *Pirates of the Caribbean: Dead Man's Chest*, when Will Turner plays against his father, Bootstrap Bill, and Davy Jones. However, it isn't just a die Turner sacrifices if he loses but a life of servitude on Davy Jones's pirate ship. For the benefit of aficionados of the game and the movie, I've provided a commentary on what actually is going on in the scene. It involves a bit more math than normal, so if it is excessive, remember that games are meant to be fun, not hard work.

Mathematical Analysis of the Game of Liar's Dice in *Pirates of the Caribbean*

If you throw n dice, then there is a clever equation that allows you to read off the chances of getting different numbers of rolls of the dice. First you need to expand the equation $(x + y)^n$. To find out the chances of getting exactly m sixes, say, in the n dice on the table, you take the term with x^m and substitute $x = \frac{1}{6}$ and $y = \frac{5}{6}$.

Let's look at the scene in *Pirates of the Caribbean* to calculate the probabilities of each claim. Fifteen dice are thrown in total by the three players. The mathematics asks you to expand $(x + y)^{15}$ to calculate probabilities.

$$(x + y)^{15} = x^{15} + 15\,x^{14}y + 105\,x^{13}y^2 + 455\,x^{12}y^3 + 1{,}365\,x^{11}y^4 +$$
$$3{,}003\,x^{10}y^5 + 5{,}005\,x^9y^6 + 6{,}435\,x^8y^7 + 6{,}435\,x^7y^8 +$$
$$5{,}005\,x^6y^9 + 3{,}003\,x^5y^{10} + 1{,}365\,x^4y^{11} + 455\,x^3y^{12} +$$
$$105\,x^2y^{13} + 15\,xy^{14} + y^{15}$$

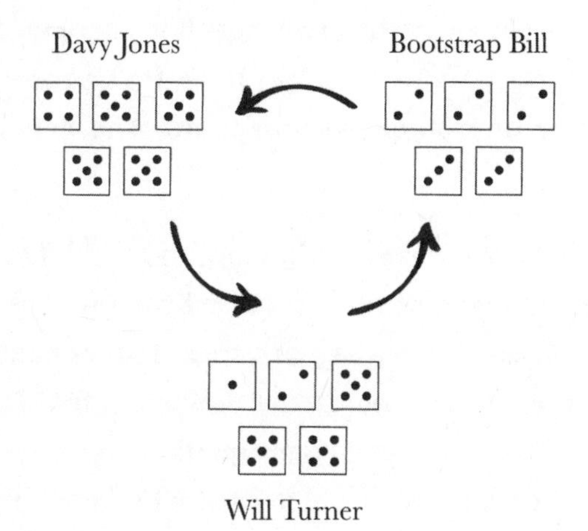

We get shown that Bootstrap Bill has thrown three twos and two threes. He starts the game by bidding three twos. For someone looking on with no knowledge of the dice, what are the chances of three twos being on the table? We look for the term with x^3y^{12}, which has coefficient 455 and calculate $455 \times \left(\frac{1}{6}\right)^3 \times \left(\frac{5}{6}\right)^{12} = 0.236$—a pretty high chance that there are exactly three twos on the table. But Bootstrap Bill is fine if there are more. The total odds come from adding up all the other terms too with x^n and $n > 3$. The chances come out at 0.468. Of course Bootstrap Bill knows there are three twos on the table because he has them.

Then Davy Jones ups the bid to four fours. For an onlooker using the mathematical equation, the chance that there are at least four fours comes out at 0.232—roughly one in four. Still pretty likely. But the chances that Davy Jones has all four fours in his hand is much more unlikely. That is got by swapping the fifteen in the equation to a five and expanding $(x + y)^5$. This comes out at roughly a one-in-three-hundred chance. On his turn Will Turner needs to start factoring in what he sees in front of him and what it sounds like his father has. He looks at his hand and sees a one, a two, and three fives. No fours! Does he challenge?

Will Turner decides not to challenge but changes the face to the fives he's holding. He bids four fives. He's counting on there being at least one other five out there. The chances of that are good.

Then Bootstrap Bill makes a big leap up to six threes. For the outsider the chances of this across all fifteen dice comes out at one in thirty-six. He can see two threes in his hand. What are the chances of another four in the remaining ten dice. Now we change the fifteen to a ten in our equation and expand $(x + y)^{10}$. The chances come out at one in fourteen. However the bets from the other two have not mentioned threes in their bids. The odds of there being four threes are probably much longer.

Davy Jones ups the ante and goes for seven fives. For the outsider the chances of this are 1 in 151. But Davy Jones is staring at four fives that he's thrown, and he's pretty sure that Will Turner has fives too, given his earlier bid. From Davy Jones's perspective, it's not a crazy bet.

This is the point where Will Turner finds himself boxed in. He's pretty sure his dad has no fives. He himself has three. Upping the ante any further would mean betting that all Davy Jones's dice are fives. The chances of that are 1 in $6^5 = 1{,}776$. But he's pretty sure that Davy Jones wouldn't have made his bet without holding lots of fives. He's trapped. He ups the bid to eight fives. Davy Jones is laughing because he knows that this is a lie. Trouble is, it's not his turn. It's Bootstrap Bill's.

Bootstrap Bill then makes the craziest bid: twelve fives. He is basically claiming that in addition to the seven that Will and Davy Jones have, he has five more fives. It's a wonderful mathematically clever line in the film. He doesn't say fifteen fives. The others could see that's impossible. But he could have been bluffing all along. So there is a chance, 1 in 1,776, that there still could be twelve fives.

Davy Jones calls him out, and indeed Bootstrap Bill loses the game and his soul to the pirate ship. But we suddenly see that by placing himself between his son and Davy Jones, he always ensured that he could sacrifice himself to save his son.

50: Pitz, the Mayan Ball Game

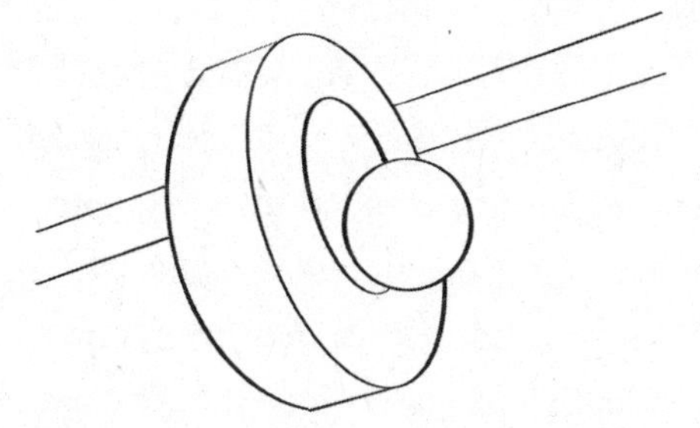

WHILE I HAVE MAINLY LEFT sports out of my journey through the world's games, I feel I can't leave South America without mentioning the ancient ball courts that were used by the Mayans to play a game they called Pitz. I've visited many of these courts on my journey around central America, including the great court at Monte Alban in Oaxaca, which I included on my visit for the Day of the Dead. The most ancient ball courts to have been uncovered date back as far as 1400 BCE on the Pacific coast near the border of Mexico and Guatemala. Access to rubber seems to have been key to kicking this game off, and examples of the rubber ball that was used have been recovered from bogs in

Mexico. The game became hugely popular, judging from the thirteen hundred ball courts that have been identified across the region.

The ball courts were either dug into the ground or surrounded by walls so spectators could look down into the court to cheer their teams on. The rubber ball was a bit larger than the size of a baseball and would have been knocked back and forth between the two teams a bit like in volleyball. The ball could be played off the walls like in squash or real tennis. Later ball courts would often have vertical stone rings on either side of the court that the teams would aim to shoot the ball through. The rings were actually quite high up on the enclosing walls, and players were only allowed to use certain parts of the bodies to strike the ball. Kicking the ball like a football wasn't allowed; nor was passing the ball with your hands like basketball. Instead players had to acrobatically strike the ball with their hips, though there were also variations where players were allowed to hit the ball with wooden sticks. An early version of hockey? Or throw in some broomsticks, and it starts to sound a bit like quidditch.

There is evidence that the game could have acted as a proxy for warfare. Rather than fighting things out on the battlefield, rival groups would agree to let the result be decided by a ball game. To be sure, there was still plenty of bloodshed, as players were known to die from the impact of a hard ball to the head. Sometimes the games would end with the losing team being ritually sacrificed on the court itself. Severed heads are commonly included in artistic depictions of the game, and they are also mentioned in connection with the game in the Mayan origin narrative *Popul Vuh*, which tells the history of the Mayan K'iche tribe.

The connection with death seems to go very deep. The courts are often considered a liminal space between the world of the living and the dead. The *Popul Vuh* features a story about a loud game that so annoys the gods of the nearby underworld that they lure the players to their deaths to stop the racket.

As my journey through South America approached its end, I expected to find more evidence of pre-Columbian games played by the indigenous communities in the region. Perhaps, as with so much else, the European conquest wiped out those games and replaced them with local variations of games imported on the invaders' boats.

On a recent trip to Colombia, I had the chance to spend the day with the Arhuaco people who live in the Sierra Nevada near Santa Marta. We wanted to explore points of connection between our two worlds, so I sat with several *mamos*, or "spiritual leaders," from the tribe to share our stories of the universe. Our discussion ranged through subjects like the origins of the universe, whether time exists, the importance of the number four, consciousness in nature, the nine-dimensional spirit world, and the symmetry and meaning of the designs used in their *mochila*, or "bags." I was fascinated by their concept of something called *kunsamu*: immutable, natural preexisting laws that will survive everyone and everything. At the time of the origin of all things, all was pure thought. For me that sounded remarkably like a description of mathematics!

Toward the end of our discussion, I asked them about the games they play. Their answer was interesting: they regard the time they spend weaving their *mochila* as a time of play. You often see children weaving the bags. From the outside, they said, that may look like child labor, but for them this is play. By the end of our time together I understood that left to themselves, they were so self-sufficient and in tune with their environment and nature that it was almost as if they had reached the utopia we are striving for. As Suits proposes in *The Grasshopper*, in utopia, where working is no longer a necessity but a choice, work can itself become a game.

But something that was even more compelling in their relationship to playing games was their concept of time. They found our concept of time totally artificial and the cause of much anxiety. They seemed to deny that we needed a concept of time. They wondered why we would be interested in playing a game to pass the time when for them there is just the current moment. Our conversation gave an interesting insight into why playing a game is contingent on a framework of time. If you take time away, can a game exist?

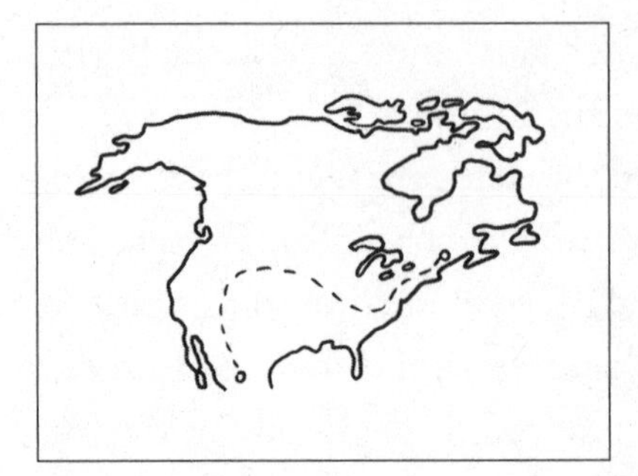

CHAPTER 12

THE UNITED STATES

CROSSING THE MEXICAN BORDER into the United States brings us to the home of some of the truly iconic games of the last hundred years. But before we open the boxes of these classics, since we are close to Nevada, let us drop in on a city that is synonymous with gaming: Las Vegas.

51: The Casino

THE WHOLE CONCEPT OF A CASINO is based on the discovery that mathematics can give you an edge when it comes to playing games of chance. Without the discoveries of Pierre de Fermat and Blaise Pascal in the second half of the seventeenth century, the modern casino would never have got off the ground. Their mathematical analysis of the throw of a die allowed the casino to rig games such that, although they might lose to punters every now and again, in the long run they will always come out on top.

That isn't to say that there were no ancestors to the contemporary Las Vegas casino, of course. The very first casino actually predates the discoveries of Pascal and Fermat. The Ridotto, Italian for "private room," was established in Venice in 1638. Although open to the public, it maintained strict rules of entry, demanding that all patrons wear a mask and a three-pointed hat. This dress code, combined with the high stakes required to play, meant that it was a casino primarily reserved for the wealthy.

The games on offer included an early forerunner of roulette called biribi. Punters could bet on any number from one to seventy, selecting from a grid of numbers rather like the table of modern roulette. But rather than spinning a wheel, the banker drew the winning number from a bag. Not quite as exciting as Caesar's Palace, but it was a start. Even before Fermat and Pascal worked out the finer points of probability, the casino had already discerned that offering a payout of sixty-four times the punter's bet would still leave the casino in the money in the long run, given there were seventy numbers to bet on.

The roulette wheel was introduced to casinos for the first time in 1796. There is some speculation that the original wheel was based on an idea that Pascal was working on already in the mid-seventeenth century: creating a perpetual-motion machine. Unsurprisingly he failed at his primary goal, but he succeeded in creating a lasting feature of the modern casino. The roulette wheel used in the casino in Paris allowed gamblers to bet on the numbers one through thirty-six and also launched the tradition of including two extra pockets for the bank, colored in green, with the numbers zero and double zero. These extra pockets gave the bank its advantage.

Bet a chip on a number, and if it comes up, you get it back with a reward of thirty-five chips. Without the zero and the double zero, that would give a fair game. But with these extra pockets, every time you bet \$1, the house is essentially making $\frac{2}{38} \times \$1$, which is about 5.2 cents. Every now and again the casino might have to make a big payout to one individual, but in the long run, thanks to the laws of probability discovered by Pascal and Fermat, it can count on coming out ahead. Every time you play, you pay.

Perhaps the most famous casino in Europe is the one in Monte Carlo. I turned twenty-one in Monaco on my way to Italy. Keen to take advantage of the occasion to do something that I couldn't have done at the age of twenty, when I learned the casino only admitted visitors who were twenty-one and over, it was settled: my twenty-first would be spent at a Monte Carlo roulette wheel. The friend I was traveling with was intrigued to know if I might have any mathematical tricks up my sleeve to give us an edge at the casino. I explained to him that casinos like the one in Monte Carlo employ too many mathematicians for us to have any hope of getting the upper hand. At best we could hope a few mathematical tricks would allow us to lose our money more slowly.

Half the numbers from one to thirty-six are colored red, and the rest are colored black. Usually if you put your money on red, and a red number comes up, you win double your stake. If you put your money on red or black and zero or double zero comes up, some casinos will apply a rule called *en prison* and pay you back half your bet. This actually means that the house odds are a little less on this bet: it's cheaper to play here—to bet on red—than anywhere else on the roulette wheel. In the long run, betting on red costs you

$$\text{(probability of losing)} \times \text{bet} - \text{(probability of winning)} \times \text{payout}$$

$$= \frac{18}{38} \times \$1 + \frac{2}{38} \times \$0.50 - \frac{18}{38} \times \$1$$

which comes out at 2.6 cents, as opposed to the 5.2 cents it costs to play anywhere else on the table. So if the casino plays *en prison*, it will take you twice as long to lose your money.

Surely, my friend said, you can do better than that. What if we just doubled our bet every time we lost? My friend appeared to be convinced that this would be a surefire way to beat the casino eventually. He was sadly unaware of the mathematical dangers of exponential growth.

What my friend was describing is a real strategy, known as a *martingale*. You begin putting one chip on red. If red comes up, you get your chip back plus another chip. If it comes up black, and you lose your original chip, then on the next round, you bet two chips on red. If it comes up red this time, you will have won your chips back plus two more, meaning you're now one chip up from where you started. If red fails to come up the second time, simply bet four chips next time. If red comes up then, you get four chips on top of your bet. Even though you've lost one chip on the first bet and two on the second, you are still left one chip up.

The way to play this system is to keep doubling your bet each time until red eventually comes up. By doubling your bet each time, when red eventually appears, you've ensured that your total winnings are always one chip more than the total money you've bet, leaving you a winner. It can sound persuasive in theory. So why doesn't it work in practice?

One problem is you would need infinite resources to guarantee a win, since there is a theoretical possibility of a run of black all night long. And even if you had a huge pile of chips, repeatedly doubling your bet can very quickly exhaust your supply. Remember how the Persian king got caught out by the power of doubling when he tried to cover the chess board with grains of rice. Most casinos maintain a maximum limit on bets precisely to stop players from exploiting this strategy. For example, with a maximum bet of one thousand chips your strategy is going to fail after ten rounds, because on the eleventh round you are going to want to bet $2^{10} = 1,024$, already more than the maximum bet.

The gambler's fallacy is to believe that if there have been eight blacks in a row, the probability must be really high of seeing red come up next. Of course, the chance of seeing eight blacks in a row is incredibly small, 1 in 256 in fact. But that won't increase the chances of getting red on the next spin of the wheel. The roulette wheel has no memory: every spin is still a fifty-fifty shot.

The casino in Monte Carlo was the scene of one of the most famous cases of the gambler's fallacy when in the summer of 1913 black came up twenty-six times in a row. So convinced were punters that red had to come up soon, after black had appeared over and over again, they threw away more and more money with every spin, falling further into the trap of believing that it must be red's turn. Given the number of roulette wheels spinning around the world, one would expect to see such a run happening every so often. After all, the odds of twenty-six blacks in a row are very close to the odds of winning the Mega Millions lottery in the United States, and enough tickets get bought that the jackpot is won about ten times a year. Needless to say, our visit to the casino on my twenty-first birthday was fun but didn't see us making enough money to move out of the student hostel we were staying in that night.

Back in Vegas, the Plaza Casino witnessed a very brave bet on red when in 2004 Londoner Ashley Revell decided to bet everything he owned on the spin of the wheel. He sold his house, his clothes, and everything he owned and took £76,840 with him to Vegas. He could have whiled away a few nights gradually losing the money, but he made the audacious decision to bet it all in one go. From a mathematical point of view, this is actually the optimal strategy. Because of the edge that the casino has, every time you play, you pay. So playing just once effectively means you pay just once.

Revell had decided to put all his money on black, but at the last moment he changed his mind. It was a good call. The ball bounced around for an age and then settled on red seven. He'd doubled his money. Rather than risking it all again, he very sensibly walked away from the casino to invest his winnings in starting an online poker company.

52: The Mansion of Happiness

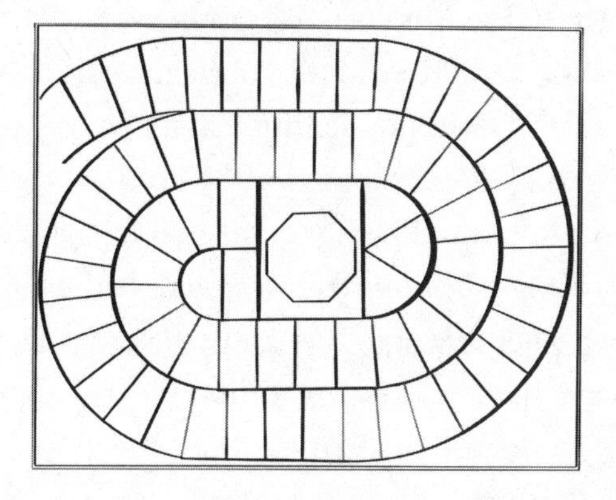

FROM THE DESERT OF LAS VEGAS, we head to America's East Coast, and more than a century back in time, for our next game. The year is 1843, and we are in Salem, Massachusetts, location of the infamous witch trials of 1692. Long gone, however, is the threat of being burned at the stake for misbehaving—now Salem is the home of a new board game that promises to guide the young on the path of virtue.

Called the Mansion of Happiness, this incredibly simple board game uses sixty-seven squares to promote Christian values. The squares spiral into the central square, which depicts an idyllic scene of men and women making music in the gardens in front of a beautiful large house, the Mansion of Happiness. Landing on a square marked "piety" advances you six additional squares, "immodesty" loses you a turn, and so forth. The set of rules accompanying the game reads as follows: "WHOEVER possesses PIETY, HONESTY, TEMPERANCE, GRATITUDE, PRUDENCE, TRUTH, CHASTITY, SINCERITY... is entitled to Advance six numbers toward the Mansion of Happiness. WHOEVER gets into a PASSION must be taken to the water and have a ducking to cool him... WHOEVER possesses AUDACITY, CRUELTY, IMMODESTY, or INGRATITUDE, must return to his former situation till his turn comes to spin again, and not even think of HAPPINESS, much less partake of it."

The game bears similarities to snakes and ladders, the game that originated in India, which also sought to become a vehicle for instilling morals in its players. The Mansion of Happiness also seems to be influenced by one of the earliest examples of board games of this type, an Italian game from the sixteenth century called Goose, which also has a spiral of sixty-three squares heading for the home square in the middle. A version of the game was actually published in the United Kingdom in 1800, but it was the puritanical atmosphere of the United States that made the game a phenomenon. The game has the claim to being the longest published game by a known author. It was on sale continuously from 1800 until its demise in 1926.

The game maker's moralism has some interesting effects on game play. For one thing, dice are not used, since they were considered immoral. (After all, the Roman soldiers had used dice to divvy up Christ's possessions at the crucifixion.) Instead, the turns were decided by the spinning of a teetotum, an ivory rod sharpened to a point at the bottom and inserted in an octagonal ivory plate marked with the numbers one to eight. Spin the teetotum to determine the number of moves your piece makes. Though dice were clearly unacceptable, the notion of the passage of life being decided by the teetotum's spin was considered to be in accordance with Protestant beliefs in predestination.

Teetotums had been used as a replacement for dice in eighteenth-century Europe as a way around a tax on dice to curb gambling, though the little-known reality was that spinning a twelve-sided teetotum wasn't actually equivalent to rolling two dice. A seven is as likely as a twelve on a teetotum but six times more likely with the throw of two dice—with potentially significant impacts on the probabilities of various games. The tax on dice could be avoided far more effectively by picking dominoes from a bag. The dominoes do replicate the throw of two dice fairly.

Taxation wasn't just restricted to dice during these times. The reason the ace of spades often has a much more florid design than any other card is that cards too were taxed. To show that the tax had been paid, the ace of spades would be stamped with an elaborate design that was difficult to forge.

The Math of Dice

Suppose you have n dice and you want to know the chances of rolling a score of m. You can use a little algebra to help. Expand the equation

$$(x^6 + x^5 + x^4 + x^3 + x^2 + x)^n$$

as powers of x, and then count how many occurrences of x^m you get. Then divide this by 6^n. For example with $n = 2$, the expanded equation becomes

$$x^{12} + 2x^{11} + 3x^{10} + 4x^9 + 5x^8 + 6x^7 + 5x^6 + 4x^5 + 3x^4 + 2x^3 + x^2.$$

So the chance of getting a score of seven is 6 in 6^2, or $\frac{1}{6}$. If you are playing Dungeons & Dragons and are using dice with f faces, then change the equation $x^6 + x^5 + x^4 + x^3 + x^2 + x$ to one with power of x up to x^f.

The spinning tops puritanical game promoters preferred to sinful dice had a secret: they too had been used for gambling since Roman times. Letters marked on the side of a four-edged spinner indicated whether you added money to the pot or took money. The word *teetotum* derives from the Latin *totum*, referring to the stake you are gambling over. You still find these spinners, called *dreidels*, used today in the Jewish festival of Hanukkah for gambling over chocolate coins.

For all its populism in the penitent atmosphere of the late nineteenth century, the Mansion of Happiness couldn't survive the twentieth-century onslaught of the new American religion: capitalism. And what better place as our next destination than Atlantic City, to track capitalism's influence on one of the most iconic games of the twentieth century.

53: Monopoly

MONOPOLY SEEMS TO HAVE a history of causing family unrest. The royal family in the United Kingdom went so far as to ban the game because of the arguments it caused. It sparked a landmark moment in my own family's folklore when I flipped the Monopoly board over in frustration that my sister was winning the game as I lost houses, money, and hotels and eventually became bankrupt. Why is it that games of Monopoly so frequently end in tears?

While my friends at school were playing on a board based on streets in London, my family's board was considered rather exotic because it featured the original streets of Atlantic City: Kentucky Avenue and Boardwalk. My parents had been given the game by American friends of theirs. On bringing the game to the United Kingdom, Waddingtons decided to anglicize it, so a company secretary was tasked with walking the streets of London to find appropriate replacements: Kentucky Avenue became the Strand; Boardwalk became Mayfair.

Part of the game's originality is in the way that players are allowed to take ownership of squares on the board. Most games up to this point treated squares on the board as free spaces on which pieces raced around. One could argue that placing a stone on a Go board means you claim that square, but the way Monopoly allows players to charge for the privilege of landing on their properties injects a fascinatingly capitalist sensibility into the world of games.

While it was my sister who amassed wealth thanks to her well-chosen purchases in our fateful holiday game, the person most famous for cashing in on the creation of Monopoly was a heating engineer from Pennsylvania called Charles Darrow. He originally brought the game to Parker Brothers in 1934, only for them to reject it for being too complicated. They cited fifty-two fundamental errors in game design as they sent him away. But after five thousand boxes that Darrow had made himself flew off the shelves in Philadelphia, they realized they'd underestimated the appeal of the game. Darrow retired a multimillionaire at age forty-seven, devoting the rest of his time to collecting orchids rather than building hotels on the streets of Atlantic City.

Shortly after Darrow's death in 1967, a case was brought in the courts against a knock-off version of Monopoly called Anti-Monopoly, in the process revealing that Darrow could only dubiously be called the original inventor of the game. It turned out that he'd put together his version after playing a very similar game one evening with friends in 1933. The true inspiration was actually a homemade game created by a Quaker from Maryland named Lizzie J. Magie.

In contrast to the modern version, which celebrates capitalism, Magie's game, patented in 1904, was designed to promote the opposite message. Called The Landlord's Game, her patent explained, its purpose "is not only to afford amusement to the players, but to illustrate to them how under the present or prevailing system of land tenure, the landlord has an advantage over other enterprises and how the single tax would discourage land speculation." The jail square was intended to be the destiny for unscrupulous landlords.

There is some interesting speculation that Magie based her design for the game on yet another original inspiration, this time an American indigenous racing game called Zohn Ahl, which a friend had introduced to her. Like Monopoly, Zohn Ahl consists of a square board with forty positions stationed around the four sides.

Instead of utilities and stations, we find creeks and rivers, which can send players' pieces back to the beginning. The design shows a definite similarity, though the game play has little to do with Magie's version.

Apparently Parker Brothers had seen Magie's early version of the game as well but had rejected it, just as it had rejected Darrow's, once again believing the rules were too complicated. But it circulated in handmade versions and over time evolved and morphed into the game that Darrow played on that fateful evening in 1933. Today it is estimated that over 250 million copies of the game have been sold with over three hundred different versions of the game, from Game of Thrones Monopoly (Boardwalk replaced by King's Landing) to a Manchester United version (Boardwalk here becomes Alex Ferguson).

The scar of losing that game to my sister stoked my obsession to master the strategy behind Monopoly. As I learned, in some ways going first gives one a little bit of an edge because it means that you have the opportunity to buy properties first. You will be hitting parts of the board before your rivals, as you will have a greater chance of being in the lead in the early part of the game. But that advantage quickly dissipates.

Since you want to own properties that people are likely to visit more often and therefore will have to pay you more rent for, my next thought was to analyze if there are squares that you land on more than others. A clever piece of mathematics that we previously explored while examining snakes and ladders can help us explore this problem. This technique, called the Markov chain, describes a series of events where the probability of the next state of the system depends only on the current state of the system, not on how it got there. In the case of Monopoly, you can be on one of forty squares, say Kentucky Avenue; the probability of the square you will land on in the next turn will depend not on how you got to Kentucky Avenue, just on the next roll of the dice.

The Math of Monopoly

To work out the chances of where you'll be next, you need to know the probability of throwing particular scores with two dice. Here is a table of those probabilities:

Dice score	2	3	4	5	6	7	8	9	10	11	12
Probability	$\frac{1}{36}$	$\frac{2}{36}$	$\frac{3}{36}$	$\frac{4}{36}$	$\frac{5}{36}$	$\frac{6}{36}$	$\frac{5}{36}$	$\frac{4}{36}$	$\frac{3}{36}$	$\frac{2}{36}$	$\frac{1}{36}$

The probability is the number of ways of getting a score divided by the total possible scores. There are six ways that one die can land. So two dice have 6 × 6 = 36 different ways. Only one of

these gives you a score of twelve: six and six, so the probability is $\frac{1}{36}$. However, there are six ways to get a score of seven: 1 + 6, 2 + 5, 3 + 4, 4 + 3, 5 + 2, and 6 + 1. Seven is the most popular score. (This is why a teetotum with twelve sides fails to replicate two dice being thrown. A score of seven on a teetotum is as likely as any other score.)

The other useful ingredient in understanding the possible state of a Markov chain is something called the *state matrix*. In the case of Monopoly, this is a 40 × 40 array where the entry in the *i*th row and *j*th column is the probability of landing on the *j*th square if you are currently on the *i*th square. For example, the entry in the twenty-first row (corresponding to being on Kentucky Avenue) and the twenty-third column (Indiana Avenue) is $\frac{1}{36}$ because that's the probability of getting a two and moving onto Indiana Avenue.

The beauty of this matrix is that if I want to know the chance of where I'll be after seven turns if I start on Kentucky Avenue, then I need to multiply this matrix together seven times to generate a new matrix whose entries tell me the probability of where I'll be next. This means I can use this matrix to explore if there are any squares that I have a higher probability of landing on than others. If I keep multiplying this matrix together, then I find that the probabilities get spread out, so that eventually every entry is $\frac{1}{40}$. No square is more likely than any other—except there are some rules and special squares that I've failed to take into account in this analysis.

For example, if I throw a double, I get to go again. Throw another double, and I get another turn. Throw three doubles in a row—an amazing feat—and I get rewarded with...being sent to jail! So from the start I could actually get as far as the (6 + 6) + (6 + 6) + (5 + 6) = 35th square, the Shortline Railroad.

There are other complications. Landing on the Chance or Community Chest squares adds another level of disruption. You could be told to "advance to Illinois Avenue," or again you might be unlucky and get sent to jail. And to add to these complications, there is actually one square on the board that never gets occupied because if you land on it, you get sent to jail. Once you are in jail, there are different ways to get out: you can pay, or hope to roll a double, or even use a "get out of jail free" card picked up from the Chance or Community Chest piles. You might even want to stay in jail because later on in the game, with a board stacked with your opponent's hotels, it makes sense to stay put. Moving around opens you up to the risk of paying high rents.

These complications already hint at an interesting bias in the game. There are many ways you can end up on the jail square. Because a player's psychology alters the probability of what they might do if they are in jail, the game no longer satisfies the rules of a Markov chain, so we need to use computer simulations to find out what's happening. When you run iterations of the game with these added rules, the jail square does indeed turn out to be two times more likely to be occupied than anywhere else. The trouble is that you can't own the jail square. But there is a way to take advantage of this information. From jail, players are more likely to throw scores of six, seven, or eight, landing them in the orange region of properties. And from there, the red properties are also more visited. These are the properties that you want to try to own.

Once you've got a set of properties, you can start to build houses on the locations. You can put from one to four houses on a property and then finally a hotel. As you increase the real estate, the rent for a player goes up if they land on your square. An interesting mathematical analysis has been done on how many turns it

takes for you to earn back your investment. It turns out that the sweet spot is three houses, and once again the orange squares are fastest in earning back the outlay.

Armed with my mathematical strategies, I am still waiting for my sister to accept my challenge of a rematch on the Monopoly board.

Personally I think that Monopoly is more than a little overrated as one of the great board games of all time. I really don't understand why so many people have it as their default game of choice to play at holiday time. I find that quite often the game just goes on too long. After a while it becomes clear who is going to win, and the rest of the game consists of this person just grinding out the bankruptcy of the other players.

In 1961 four students at the University of Pittsburgh found their game going on for so long that they thought they were on course to record the longest game of Monopoly in history. After two days of playing, they called the local news station, which sent a crew to record the epic game. On the third day it became clear that the bank was about to run out of money.

They weren't quite sure what to do if this happened, so they contacted Parker Brothers to find out how to proceed. The president of Parker Brothers wired back a response: "Refuse to let bank fail. Rushing one million Monopoly dollars to you by airmail—carry on." Sure enough a million dollars in toy money was flown to Pittsburgh and forwarded to the students via an armored truck.

The game was allowed to continue. But after five days, the game seemed to be going in circles. The students lost the will to play. They stopped the game and added up their assets to determine a winner. Frankly I feel this way when I play the game for just two hours. That's why I was quite excited to discover a rather beautiful analysis of the shortest way to win the game, which Dan Myers, a sociology professor at the University of Notre Dame, and his son came up with.

In a two-player version of the game, if the dice roll nicely and the Community Chest and Chance cards are arranged carefully, it is technically possible to win the game in twenty-one seconds. Both players start with $1,500.

Player 1, Turn 1: Roll: 6-6. Lands on: Electric Company. Action: None. Double means roll again. Roll: 6-6. Lands on: Illinois Avenue. Action: None. Another double, therefore roll again. Roll: 4-5. Lands on: Community Chest. "Bank error in your favor, Collect $200." Action: Collects $200 (now has $1,700).

Player 2, Turn 1: Roll: 2-2. Lands on: Income Tax. Action: Pay $200 (now has $1,300). Double means roll again. Roll: 5-6. Lands on: Pennsylvania Railroad. Action: None.

Player 1, Turn 2: Roll: 2-2. Lands on: Park Place. Action: Purchase ($350, now has $1,350). Double means roll again. Roll: 1-1. Lands on: Boardwalk. Action: Purchase ($400, now has $950). Another double, therefore rolls again. Roll: 3-1. Lands on Baltic Avenue. Action: Collect $200 for passing Go (now has $1,150). Purchase three houses for Boardwalk, two for Park Place ($1,000, now has $150).

Player 2, Turn 2: Roll: 3-4. Lands on: Chance: "Advance to Boardwalk." Action: Advance to Boardwalk. Rent is $1,400, only has $1,300. Bankrupt. Game over.

The chances of this ever happening have been calculated as 1 in 271 trillion, however, so don't hold your breath for such a swift end to your next game of Monopoly.

Many people feel that games are too biased toward people who are good at probability and logic and don't favor the more literary types. When my next game appeared on the scene, many bookworms felt that they finally had a game where they had an edge over the mathematicians. But, as we shall see, there was more going on behind the words than they may have realized.

54: Scrabble

"WE SHOULD DOWNGRADE THE SCORE of a Z from 10 to 6." In January 2013 the media whipped up a mini storm in the gaming community by broadcasting research that suggested that the values of letters in Scrabble were misaligned with their current use in the English language.

Joshua Lewis, a cognitive scientist at the University of California, San Diego, had written a computer program to recalibrate the value of letters in the game of Scrabble, given that the dictionary of acceptable words had changed since the game was first invented seventy-five years ago. For example, QI (meaning Chinese life force) was introduced as an acceptable word in 2006, which made the highly tricky letter Q much easier to use. Lewis suspected that all the changes to the dictionary of Scrabble words should indeed affect the value of each letter.

His computer program suggested that ten tiles should be reduced in value. Both X (eight points) and Z (ten points), for example, should now be valued at six points, given how many more two- and three-letter words were now available using these letters. Lewis's analysis suggested that four other letters should increase in value. For example, V is only valued at four points in the current game, but it's really worth five points in the current lexicon of Scrabble words. Only twelve letters were considered to be acceptably valued by the algorithm.

Although the media got very excited by this challenge to the game, Scrabble players were less enthusiastic. They felt that Lewis's analysis missed the point of the game. The anomalies in the way letters were valued formed part of the game play. The overvaluing of Z was well known and exploited by players. Flattening out scoring, as Lewis suggested, would make the game "fair" but less fun, eliminating the possibility of the exciting dramatic swings that can make the game so exhilarating. As Stefan Fatsis, author of *Word Freak*, wrote, "The consensus of my math-brained Scrabble colleagues is that this would be like a dose of lithium for the game."

To be sure, when Alfred Mosher Butts, the creator of Scrabble, first assigned values to letters in 1938, his first approach to the problem was indeed to determine the frequency of letters in the English language. This approach to language had already been exploited many centuries before, not in the context of a game but as a key to code cracking.

One of the most ancient codes, which goes back to Caesar's time, is the substitution cipher. This code systematically exchanges letters for alternative letters. For example, every instance of A gets replaced by another letter, say G. Then B is replaced by one of the remaining letters. In this way each letter of the alphabet gets assigned a new letter. There are a lot of different codes one can choose. There are $26! = (1 \times 2 \times 3 \ldots \times 26)$ different ways to rearrange the letters of the alphabet.

If a hacker intercepts a message, then at first sight it seems an impossible task to decode the text. After all, don't they need to try all these different permutations? Testing one per second would take beyond the current lifetime of the universe. But there is a weakness to this crypto system that the ninth-century polymath Ya'qub al-Kindi spotted, and it relates to how letters are scored in Scrabble: some letters appear more frequently than others. For instance, in the English language E is the most popular letter to appear in any text, occurring 13 percent of the time, followed by the letter T, which crops up 9 percent of the time. Letters also have their own personalities, which get reflected in other letters they like to be associated with. Q is typically followed by U, for example.

Letter	Frequency	Letter	Frequency
E	12.7	M	2.4
T	9.1	W	2.4
A	8.2	F	2.2
O	7.5	G	2.0
I	7.0	Y	2.0
N	6.7	P	1.9
S	6.3	B	1.5
H	6.1	V	1.0

R	6.0	K	0.8
D	4.3	J	0.15
L	4.0	X	0.15
C	2.8	Q	0.1
U	2.8	Z	0.07

Butts was unaware of this work on letter frequency done centuries earlier and instead undertook his own research. Using copies of the *New York Times*, the *New York Herald Tribune*, and the *Saturday Evening Post*, he made copious records of the frequencies of letters used in the articles in order to inform his decision about the game he was creating. The first version of the game had no board and no scores for different letters. Instead he conducted his statistical analysis of letter frequency to determine how many of each letter there should be in the game. He settled on one hundred letters, with rare letters like Q, Z, or X only occurring once, while letters like E and T had multiple copies, mirroring the letter frequency he'd detected in his analysis of the newspapers. Given that K only occurred once, you'd never see the word KINK on the board. These decisions ended up excluding swathes of the dictionary.

He decided, however, to cut down the number of S's, as these were too easy to add to the ends of words to make them plural. You only find four S's in a Scrabble set. As happens with all games, it was only when he got friends playing the game that he realized what his final decisions about letter frequency and scores should be.

You might think that of all the games in this book, Scrabble would favor the wordsmith over the mathematician. But you'd be wrong. The best Scrabble players treat playing the game like a puzzle, seeking out clever ways of inserting a few letters that can be read in multiple directions, ratcheting up the points. As Robin Pollock Daniel, one of the world's highest-rated Scrabble players, told the *Toronto Star*, "Words are involved but to me at least it's more about math. Scrabble has just so many dimensions. It's visual. It's spatial. The game's inventor, American Alfred Mosher Butts, was an architect." John Williams, a former director of the United States' National Scrabble Association, agreed: "At the highest level Scrabble is a math game. It's like poker. It's all about probabilities and managing a rack [of tiles]."

Good players of the game like Daniel commit to memory the array of two-letter words that can be used in cunning ways to squeeze in high-scoring combinations. There are 107 officially acceptable two-letter words. See how many of the following you can define: AA or GU or ZO. The first is rough, jagged lava found in Hawaiian volcanoes. The second is a sort of violin made in Scotland. The third is a Himalayan cross between a cow and a yak. None of these are words I would anticipate bringing into everyday discourse anytime soon.

The other strategy is to learn all the seven- and eight-letter words that contain a good smattering of the common letters like A, E, I, N, R, S, and T—for example, TAURINES (amino acids containing sulfur and important in the metabolism of fats) or PINASTER (a pine tree with long, thick needles and clustered cones, native to the coasts of the Mediterranean and Iberia). Using all the letters on your rack earns you a fifty-point bonus called a *bingo*. Having this esoteric list of words at your fingertips is another useful stratagem. Although you only have seven letters on your rack at a time, generally you'll need to use a letter on the board to connect your word to, which is why knowing the eight-letter words is valuable.

The highest recorded score for a single turn in Scrabble dates back to April 1982 when Dr. Khoshnaw scored 392 points for CAZIQUES getting the Q on a double letter score and the C and S on triple word scores, as well as scoring fifty points for using all seven letters in his rack. CAZIQUES means "West Indian chiefs."

Being an eloquent and articulate English speaker may end up a disadvantage in playing the game. Such players get obsessed with cooking up beautiful words from their rack of letters rather than strategically analyzing the board for triple letter squares that might add to their score. After all, it is a number rather than a word that ultimately determines the winner of the game. A telling indication of the handicap that being an eloquent English speaker might pose is that some of the best players of the game hail from Thailand. The 2003 Scrabble World

Championship saw two Thai players competing in the final game for the title, which was eventually won by Panupol Sujjayakorn, who admitted he barely had conversational English at the time.

As if to rub it in that you really don't need to speak the language to win, one of the longest running champions of Scrabble in the English language, New Zealander Nigel Richards, applied his strategy to the French language and in 2015 beat all the native French speakers at the game to become the French Scrabble champion despite not speaking a word of French.

At first sight it looks like Scrabble might be a degree harder in some languages. The official Scrabble dictionary in English lists two hundred thousand acceptable words. French has nearly double that at 386,000 words. Top of the list is Italian, with more than 661,000 words. Given that English is meant to be such a rich, complex language, how come its Scrabble dictionary is so much smaller than those of these other European languages? The key is that French and Italian are examples of languages where words can have multiple forms, given, for example, that verbs can be conjugated. Remember learning in all those verb endings in French lessons? In French each root has, on average, six versions of the word that can be used in the game. Italian is even richer in variations. Using just 6,196 verb stems, you can generate 319,000 words that are acceptable in a game of Scrabble.

The way that computers play the game can give us further insight into the underlying workings of Scrabble. What emerges is that making the highest-scoring word out of your available letters isn't always the best strategy. Rather, it is more important that there is a high probability that the letters a player is left with have a good chance to create a good word when combined with the new letters drawn to replace the ones just used. You also don't want to make a move that opens up the board for your opponent to exploit a high-scoring square. With the need to analyze a tree of forward moves and to keep track of the letters that have already been used and what might be still available, the game starts to look like poker rather than a gentle word game. Indeed, a mathematical analysis of the game play has revealed that it actually has the character of a puzzle like the Traveling Salesman Problem, where there is no known fast algorithm to pick out the optimal play.

The challenge of the Traveling Salesman Problem is to come up with a strategy that finds the shortest path around a network of cities connected by roads of different lengths. We haven't found a smarter strategy than trying all the possible paths and picking the shortest. The same problem is faced in trying to find optimal strategies to play many of the games in this book.

This is why when a game is solved and an optimal game play is uncovered, it is often done by a similar brute-force analysis of considering all possible games.

Where Scrabble rewards knowing your seven- and eight-letter words, we certainly can't overlook another American word game that became a latter-day phenomenon by zeroing in on five-letter words.

55: Wordle

JOSH WARDLE CREATED WORDLE as a puzzle for his partner. The Covid-19 pandemic was raging, and the couple were keen for any distractions from the news.

The way the game works is to challenge players to guess a five-letter word. You have six guesses to identify the word, and on each guess, Wordle will tell you if you've got the right letter in the right spot, by coloring the letter's square green. A correct letter in the wrong spot shows up on a yellow background. A letter that isn't in the word in any spot shows up on grey. There is only one word a day. Everyone gets the same word. Wardle compares the charm of Wordle to eating a croissant at breakfast. One is really tasty. More than one, and you begin to feel sick.

You might argue that Wordle isn't a game but a puzzle. And I'd agree with you. But the interesting thing is that Wordle took off as a global phenomenon when Wardle added a feature that upped the gaming element. Allowing players to share their scores suddenly enabled them to compete against each other for how quickly they got that day's word. If you think about it, quite a few games partake of the excitement of multiple players simultaneously doing the same puzzle. Even Scrabble has an element of that, especially when people simply work to make the best words with their letters rather than playing strategically to block their opponent's next move. The gaming element has been dialed up further by the creation of WordleBot, which even grades your play with a score from zero to ninety-nine based on how skillful or lucky you were in your solution.

Wordle's core gaming mechanic isn't exactly unique. My sister and I used to play endless games of something called Mastermind, where one player would choose four colored pegs that remained hidden, and the other player would try to guess the sequence. Each guess would be scored with a black peg indicating a correct color in the correct place and a white peg to indicates a correct color in the wrong place. All Wardle has done with this formula is to replace the sequence of colored pegs with a five-letter word.

Given that Wordle is about words, this is another game that literary types might think they have an advantage in playing. However it wasn't too long before mathematicians started doing an analysis of whether there are strategies to minimize the number of guesses needed to find the five-letter word. Given that we are trying to crack a code, the ideas that Ya'qub al-Kindi came up with to analyze letter frequencies will be useful here too. As we saw above, the letters E, T, and A are the most commonly found in words in the English language. So starting with a word that uses these is likely to get you some useful information.

Professor Barry Smyth, a computer scientist at University College Dublin, found that if you just want to use a single starting word, then TALES is your best bet. Using this word leads to success in over 95 percent of games, with an average game length of 3.66 rounds, and 83 percent of these games are completed within four rounds.

An even better strategy is to use your first two turns to gather information about ten letters. He found that trying CONES followed by TRIAL leads to success in just over 96 percent of games with an average game length of 3.68 rounds, and 86 percent of games are completed within four rounds.

His simulations reveal that Wordle strategies generally fail to solve about 4 percent of the five-letter words possible. The killer words generally have repeated letters. For example HITCH catches out many strategies to sniff out the word in six guesses.

Another interesting result of his analysis is the finding that the first and fourth letters are the hardest to pin down. The easiest is the second letter; 72 percent of the time the second letter has been guessed successfully by the fifth round.

Another game where a player's moves try to whittle down the choice of a word is twenty questions, where your opponent selects the answer from the Oxford English Dictionary, and you have to ask yes or no questions until you find it. Most opt for questions that elicit properties of the word: "Is it alive?" and "Can I eat it?" But mathematics can guarantee a way of getting the answer in twenty questions. Make a list of all possible answers in alphabetic order. Then ask, "Is the answer in the first or second half of the list?" Now remove the half that doesn't contain the word and ask again, "Is the answer in the first or second half of the new list?" This surefire, if boring, strategy will identify a single word in a list of $2^{20} = 1{,}048{,}576$ words. That's over a million different words. The Oxford English Dictionary contains only 273,000 words.

The Math of Wordle

Claude Shannon initiated the mathematics of information theory in a seminal paper published in 1948. He produced a formula that quantifies the amount of information you can glean from a message you receive, called the message's entropy. We can apply his formula to analyzing Wordle. Consider a version of Wordle where you are trying to guess a two-letter word from a list of seven possible words: ON, IN, NO, NE, OW, SO, AD. If I enter the word ON as my guess, then the feedback I get will allow me to uniquely identify the answer next time. This guess gets me a lot

of information. It is a guess with high entropy. Each score only corresponds to one possible answer. But guess the word AD, and there is a chance that the score still has six possible answers. This guess will elicit less information and is a guess with low entropy.

Wordle Score	**Guessing *ON***	**Guessing *AD***
Green Green	ON	AD
Green Yellow		
Green Grey	OW	
Yellow Green		
Yellow Yellow	NO	
Yellow Grey	NE	
Grey Green	IN	
Grey Yellow	SO	
Grey Grey	AD	ON, OW, NO, NE, SO, IN

$$\text{Entropy} = \sum_{\text{All events } i} -p_i \log_2 p_i$$

$$\text{Entropy of ON} = 7\times\left(-\frac{1}{7}\times\log_2\frac{1}{7}\right) = 2.8$$

$$\text{Entropy of AD} = \left(-\frac{1}{7}\times\log_2\frac{1}{7}\right)+\left(-\frac{6}{7}\times\log_2\frac{6}{7}\right) = 0.4916$$

The best starting words are those that have high entropy and yield as much information as possible. In Wordle each of the five letters gets a score of green, yellow, or grey, which means 3^5 = 243 different scores, although 5 of these aren't possible. You can't have four greens and one yellow. There are approximately thirteen thousand acceptable five-letter words. The guess with high entropy is one that evenly distributes these words across all 238 scores, assigning roughly fifty-five words to each. It's a bit like the tactic in twenty questions where each time you divide the words left equally in two. Any deviation from this loses possible information.

Wordle isn't just good fun—its social component actually ended up saving Denyse Holt's life. The eighty-year-old retired teacher from Illinois was religious about swapping her Wordle score each morning with her daughters on the West Coast of the United States. Then one day no score arrived. After twenty hours of silence, the daughters began to get worried. No one was answering the landline, so they got a neighbor to go around and check on their mum. They discovered a man suffering a mental health crisis had entered Holt's house and locked her in the basement. If not for Wordle, there's no telling how long she might have been there. Her addiction to the game ended up saving Holt's life.

The success of Wordle can be judged in part by the extraordinary range of variants it has spawned: Squardle, six games of Wordle played simultaneously on a 5 × 5 grid; Quordle, a kind of quantum Wordle played in four different universes where each guess is applied to all four puzzles; Worldle, identify a country from its outline; Nerdle, six attempts to guess the correct mathematical calculation. You could waste the whole morning puzzling away if you're not careful.

From the recent social phenomenon of Wordle, we head next to the city of Philadelphia, where we will explore another older but still very social game. Whereas Wordle is about building connections, this one boils down to conflict resolution.

56: Rock Paper Scissors

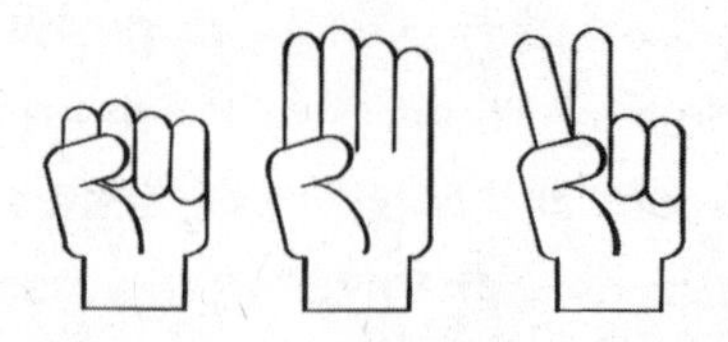

JIANDAO SHÍTOU BU IN CHINA. *Jan-ken-pon* in Japan. *Ro-sham-bo* in California. *Kai-bai-bo* in Korea. *Ching-chong-cha* in South Africa. The game of Rock Paper Scissors (RPS) is played all around the world to settle disputes from the playground to the boardroom. Two auction houses, Sotheby's and Christie's, famously settled a dispute over the right to auction a collection of Impressionist paintings by Paul Cézanne and Vincent van Gogh with a single game of Rock Paper Scissors.

After a count of three, each player chooses between three different hand positions. Rock is a closed fist. Paper is a flat hand. Scissors is a V shape made with two fingers. Scissors beat paper: after all they can cut the paper to shreds. Rock beats scissors: after all it can blunt the scissors. And finally paper beats rock because... well, just because.

In reality, a piece of paper isn't much use against someone hurling a rock at you, but the calm assertion of paper's power to defeat a rock hints at the Chinese origins of the game. In ancient times a petition to the Chinese emperor was symbolized by a rock. The emperor would indicate whether he'd accepted the petition by placing a piece of paper above or below the rock. If the rock was covered by the paper, the petition was refused and the petitioner defeated.

As a mathematician I find this game fascinating because winning depends on doing what mathematicians do best: spotting patterns. If you can predict what your opponent is going to do next, then you're in—and the best way to do that is by catching onto the pattern of behavior they've established. So when I was in the United States a few years ago, I decided to put my pattern-searching abilities to the test and enter the famous Rock Paper Scissors League Championship that is held each week in the Raven Lounge in downtown Philadelphia.

Each competitor needs their own individual RPS name to enter the league. Regulars include Paper Tiger, Slanted Scissors, and Silly Putty. I decided to try some academic intimidation and plumped for The Professor. A referee stepped up to the stage, dressed as if ready to take charge of a basketball game, and reminded everyone of the rules. "You need to win twice to win a round. It's best of three rounds. There will be penalties for anyone drawing late, anyone shooting a vertical paper. Any use of Dynamite, Bird, Well, Spock, Water, Match, Fire, God, Lightning, Bomb, Texas Longhorn, or other nonsanctioned throws, will result in automatic disqualification." With that, contestants were invited to take their turns competing in the knockout competition.

After watching a few furiously fought rounds, suddenly I heard someone call my name. I turned to face someone called Zombie Llama, a strange man with hands covered in what I hoped was fake blood and a gas mask. This was clearly a strategy to stop me looking into his eyes and guessing his next move—though I must admit the sight of Zombie Llama succeeded in making me feel quite nervous. Nevertheless, the crowd's chanting of "The Professor! The Professor!" as I ascended the stage steeled my nerve.

As something of a novice compared to the regulars at the Raven Lounge, I decided to go for a strategy that would give my opponent no chance to spot a pattern in my behavior. Behaving randomly is actually surprisingly difficult for humans. So I tapped into a bit of mathematics that is totally random and devoid of patterns: the number pi.

Pi is the number you get when you take the circumference of a circle and divide it by the diameter. It is a number that people have been calculating ever since we've been doing mathematics. It begins 3.1415926535...and then spirals off to infinity in a string of digits with seemingly no rhyme or reason to predict what digit comes next. Indeed it is conjectured that the decimal expansion behaves as if a ten-sided die were deciding the digits. Throw such a die infinitely often, and at some point it will generate any sequence of numbers you want. This implies that pi too will contain any string of digits, even your bank account details, somewhere in its expansion.

To translate pi into a strategy for Rock Paper Scissors, I decided to use a simple algorithm. I would run through the decimal expansion using each digit to make my decision about what to throw next. One, two, and three corresponded to a rock; four, five, and six to paper, and seven, eight, and nine to scissors. So 3.14159 became rock, rock, paper, rock, paper, scissors. You might wonder what to do when you hit a zero. Curiously zero doesn't appear in the expansion of pi until the thirty-second decimal place. If I managed to make it that far through the tournament, I'd allow myself a free choice of throw.

Amazingly my strategy worked. It only gave me a fifty-fifty chance of winning, but I thought that might be sufficient while facing off against hardened competitors like Zombie Llama who came to the Raven

Lounge every week to throw rocks and scissors. My strategy guaranteed that my opponent would have no clue as to what I was going to throw next. The coin landed in my favor, and I was through to the next round, where pi worked again, getting me to the quarter finals.

In the quarter finals I met my match: Dick Nasty. After some tight draws, he seemed to tap into what I was doing.

Had he really spotted the use of the digits in pi or was it just luck? When his paper finally conquered my rock and knocked me out, he offered his hand to commiserate on my exit from the competition. It was then that I noticed a rather curious tattoo on his arm: a collection of squares of different sizes corresponding to the numbers 1, 1, 2, 3, 5, 8, 13 with a Fibonacci spiral traced through them. The Fibonacci numbers are intimately related to the golden ratio, which like pi has an infinite decimal expansion that can be used to randomize your choices. My pi had been trumped by Fibonacci. If I had to be knocked out of the competition, it was no dishonor to lose to someone who was prepared to mark his body permanently with the wonders of mathematics.

Rock Paper Scissors is an interesting example of a mathematical phenomenon called *nontransitivity*. If *a* is bigger than *b*, and *b* is bigger than *c*, then we can infer that *a* is bigger than *c*. But this logic does not work when you apply it to Rock Paper Scissors. Rock beats scissors. Scissors beats paper. But that doesn't mean rock beats paper.

There are some very interesting and counterintuitive examples of nontransitivity in games. I have three dice colored red, yellow, and blue. If I told you that the dice are loaded so that, on average, red beats yellow and yellow beats blue, most people's intuition would be that red would on average beat blue. However, it is possible to rig the dice so that the opposite is true.

The Math of Nontransitive Dice

The red die has five sides with the number three and one side with the number six.

The yellow die has three sides with the number two and three sides with the number five.

The blue die has five sides with the number four and one side with the number one.

If I throw any two of these dice, then there are thirty-six different ways they can land. With the red and yellow dice, fifteen of these have (R, Y) = (3, 2), with a win for red; three have (R, Y) = (6, 2), with a win for red; three have (R, Y) = (6, 5), with a win for red. That's a total of twenty-one out of thirty-six ways red wins. So there is an advantage to use the red die.

Go through the same analysis for the yellow and blue, and you find again twenty-one out of thirty-six ways the yellow beats the blue. But now the strange thing is that comparing the blue against the red, there are twenty-five ways out of thirty-six that the blue wins.

My three colored dice can be extremely useful in a competition. You show your opponent the dice and let them choose whichever they think is the most powerful. But once they have chosen, there is always another die that on average will beat the one chosen by your opponent.

Nontransitivity is often quite important in making an effective game. For example, in the role-playing game Dungeons & Dragons, each character is skillful at some things, perhaps fighting, but not so adept at others, like casting spells. The important quality is setting up a varied range of characters where each will be able to beat some monsters but ineffectual when encountering others. A good team has a balance of all traits so that, as with the dice, you can pick the right character to compete in any particular context.

One interesting game that exploits nontransitivity is called Penney's game, which involves tossing coins. To play, you ask your opponent to select a sequence of three choices of heads or tails. You then choose your own sequence. The person whose sequence occurs first as you continue tossing the coin wins. The rather surprising quirk is that whatever sequence your opponent chooses, there is always another sequence of three coins that is more likely to occur than their choice.

First Player's Choice	Second Player's Choice	Odds in Favor of Second Player
H**H**H	**T**HH	7 to 1
H**H**T	**T**HH	3 to 1
H**T**H	**H**HT	2 to 1
H**T**T	**H**HT	2 to 1
T**H**H	**T**TH	2 to 1
T**H**T	**T**TH	2 to 1
T**T**H	**H**TT	3 to 1
T**T**T	**H**TT	7 to 1

You basically choose the first two choices of your opponent and stick in front of that the opposite of their second choice.

Before we take leave of our journey through American games, there is one more stop: a game that underscores how important mathematics can be to the dynamics that make playing games so thrilling.

57: Ticket to Ride

WHEN THE TIME COMES for Phileas Fogg to cross the American continent, he does so primarily by train. Starting at Oakland train station in California, he takes a winding path across the states, with a detour via Salt Lake City to explore Mormonism. While rescuing his companion Passepartout from a group of Sioux who attack the train, he manages to miss his connection and is forced to catch up with the train via a wind-powered sled. They are then almost stalled by damage to the bridge across Medicine Bow River in Wyoming, but the train driver agrees to attempt a crossing in spite of the danger. By taking the bridge at full speed, he manages to just make it to the other side before the bridge collapses into the river. They finally arrive in New York after seven action-packed days of traveling.

Today it is virtually impossible to repeat the same overland journey in the same time. Flying has all but destroyed consumer rail as a viable option for crossing the continental United States. Even so, Phileas Fogg's train journey across America has given rise to one of the most popular board games of recent years, Ticket to Ride. As the back of the box announces,

> October 2, 1900—28 years to the day that noted London eccentric, Phileas Fogg accepted and then won a £20,000 bet that he could travel "Around the World in 80 Days." Now at the dawn of the century it is time for a new impossible journey. Some old friends have gathered to celebrate Fogg's impetuous and lucrative gamble—and to propose a new wager of their own. The stakes: $1 million in a winner-takes-all competition. The objective: to see which of them can travel by rail to the most cities in North America—in just 7 days. The journey begins immediately...

First published in 2004, this game has become one of my family's favorites, and it won many awards when it first came out. Alan R. Moon recognized that he had designed a winner when, after finishing playing a prototype with friends, he felt a pain in his leg. On retiring to the bathroom, he discovered that a huge splinter had entered his leg and remerged some four inches further down his thigh. "I told my play-testers I had to go to the hospital to get the splinter out. They seemed unfazed and began another game as I left, and they were still playing when I got back. I kept the splinter as a souvenir, since most of it came out in one piece!" If the game could distract him from such a trauma, then it had to be good. Once it was published, the public agreed, and it went on to sales of over eight million copies worldwide.

For my family to truly embrace a game, it is important, above all else, that the rules must be simple. I quite like games with intricate rules, but the rest of the family glaze over if it takes longer than a couple of minutes for me to bring them up to speed. This game they got.

The board depicts a map of America with a network of rail lines connecting cities. The aim of the game is to claim the railroads. Each player gets cards with specific cities, which they are tasked with trying to connect with their company's tracks, represented by trains of the color they've chosen. These cards are the "tickets" of the title of the game. You get dealt three possible tickets at the beginning of the game, and you can accept all three challenges or discard one. During the game you can add to your stack of tickets by picking up three more and choosing at least one to add to your challenges. You score points for each route completed, but if you get blocked by other players, you will need to find an alternative, longer route to your destination.

The Math of Ticket to Ride

In addition to the challenge of completing your tickets, every time you connect two adjacent cities on the map, you score points. The distance between cities varies from one to six train lengths. One reason I love this game is that the designer of the game has employed a very interesting gearing mechanism to reward risky strategies, patience, and uncertainty. Look at the points you score for connecting two cities of different lengths.

Number of Trains to Complete Route	Points Scored	Value of Each Train
1	1	1
2	2	1
3	4	$1\frac{1}{3}$
4	7	$1\frac{3}{4}$
5	10	2
6	15	$2\frac{1}{2}$

For example, claiming Kansas City to Omaha requires laying down one train and scores you one point. On the other hand, lay down six trains to claim the link from Portland to Salt Lake City, and you score fifteen points. It might take you longer to collect the cards to play those six trains, but your patience is rewarded by each train scoring you 2.5 points apiece. Also you can lay these six trains in one turn, whereas it would take six turns to claim six links of length one. In this way, the game is cleverly geared to incentivize risk taking. At the same time, your gamble to collect the cards to play six trains is always at risk of failure.

As the length between the adjacent cities increases, the point value increases, as in the following diagram.

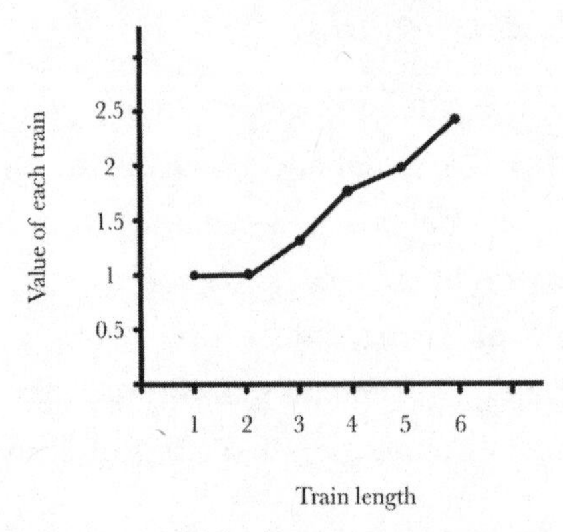

The points you earn for completing a route simply correspond to the smallest number of trains it takes to get from one city to the other. For example, the route from Seattle to New York requires a minimum of twenty-two trains to be laid, and you score twenty-two points if you succeed. You lose twenty-two points if you fail to complete it by the end of the game, so be careful how many challenges you pick up. The other tactic that increases the value of the points you earn for placing a train is to try to pick tickets where part of the tracks claimed can be used for multiple tickets.

For example, if I have the ticket for Seattle to New York, which needs twenty-two trains, and Vancouver to Montreal, which needs twenty trains, then because of the huge overlap in the routes, I can claim both routes with the same twenty-two trains. This means one train is scoring two points rather than just one if I only had a single ticket. Looking for overlaps in the routes you choose will increase the value of the trains you are playing. To weaken the score of your opponent, you can try to block the routes it seems they are going for, forcing them the long way around, and hence using more trains to gain the same number of points as won the short way. The value of the trains goes down.

Another interesting aspect of the game design is the mathematical network that represents how all the cities are connected. The different ways of getting from one city to another mean that the challenge of connecting cities set by the ticket cards can vary in difficulty. A fun way to think about this is to consider the network as an electrical circuit and to determine the routes of least resistance. More routes from points A to B reduce the difficulty. Match this against the point score, and one can start to pick out the more optimal tickets. For example the network rewards going from Seattle to New York but penalizes trying to go for the shorter route of Kansas City to Houston.

Overall, given the scoring and the probability of achieving certain goals, a mathematical analysis of the game reveals that longer routes are overvalued, which provides a strategy for anyone playing the game to gain an edge.

The mathematics of point scoring is an absolutely integral part of Ticket to Ride's game design. If the designers had gotten the gearing wrong, the game would fall flat. Getting it right means maintaining every player's interest right up to the moment the last train is played, even to the extent of not noticing a huge splinter in your leg.

My own train has finally come to the end of its American journey. I have reached New York, ready to make the last sea crossing of the book as we make our way to our final destinations of Africa and Europe.

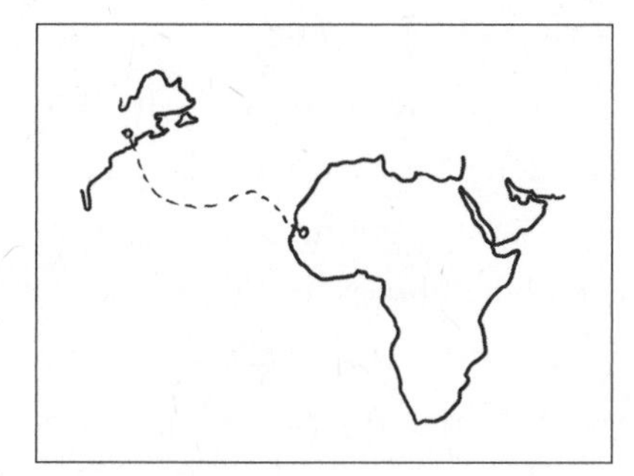

CHAPTER 13

THE ATLANTIC OCEAN

Computer Games

CROSSING THE ATLANTIC BY boat is one of the few sea passages that I have actually made in real life. It took six days, although the captain admitted to me that he could have done it in three, but then the punters on board wouldn't have got value for money. They were there to enjoy the ride, not get to their destination. Not me: I was working as part of the entertainment for the paying customers.

Cruises have always seemed a strange kind of holiday to me. I get the point of a cruise if you are visiting interesting islands on the way. But this particular voyage just consisted of one long journey from one side of the Atlantic to the other without much distraction. It was like being stuck in a huge hotel unable to go out to explore the lay of the land beyond the perimeter. With nowhere to go, the likes of me were recruited to keep things interesting.

I was giving lectures on board about great unsolved mathematical problems. The Riemann hypothesis. The Poincaré conjecture. NPvP. The trouble was that a couple of the people on board were convinced they had solutions to these mathematical conundrums. I spent most of the cruise trying to find a safe place where they couldn't find me and harangue me with their crazy theories. As I hid away in the hold, my Nintendo DS and Zelda proved a helpful way to while away the tedious hours at sea. So what better subject for our final sea interlude than video games themselves? The best video games can be like an engrossing novel. Sometimes even better, since you get to be a character in that novel. Still, over the years I have had to be quite careful about my relationship with video games. My free time is short, and yet the addictive quality of some of these worlds can find me losing hours immersed in these alternative universes. The pages that follow explore how video games work and how they relate to the more classic game types that have taken up the bulk of our journey thus far.

58: Prince of Persia

THE FIRST TITLE THAT MADE me realize video games' potential to consume huge swathes of my life was Prince of Persia. I discovered this game in 1991 while I was a postdoc in Jerusalem, perhaps an appropriate place to find myself venturing through dungeons battling the grand vizier on my way to freeing a princess.

When I first arrived in Jerusalem, I was pretty lonely. It was my first time living abroad for a long period. Everything felt very alien. Losing myself in this game for a few hours at the end of the day was a useful escape. The game was inspired by a mix of *Raiders of the Lost Ark* and *A Thousand and One Nights*, and I played my copy on one of the Apple II computers that the department owned. One thing I don't like about playing video games is the rather antisocial nature of playing on your own. It doesn't tick that psychological need for relatedness that playing games with others provides. But because I played in the department, a number of other people started to get intrigued by the challenges facing this tiny white-clad figure dancing around the screen.

After a while I found I was sitting regularly with Rama, one of the secretaries in the department, trying to battle our way through the different levels in the dungeons. Rama was fantastic at fighting. Perhaps her time serving in the Israeli army gave her an edge. On the other hand, I was good at solving the puzzles we faced. It felt like we were on an adventure together, and together we made a pretty good team. We'd be stuck, unable to clear a level, and the next day one of us would come up with a new idea.

I still remember the excitement of jumping through a mirror in one of the rooms and seeing a ghostlike doppelganger emerge on the other side and run off. I think it took us several months to go through twelve levels until we eventually met up again with that mysterious doppelganger. Did we fight ourselves? Was there a way to become one again? Finally we managed to defeat the grand vizier and release the princess. The satisfaction was like putting the final QED at the end of a proof.

Thinking about how video games work as games can lead to strange conclusions. No doubt Prince of Persia engaged me like the best games. The whole adventure was totally addictive, and at the height of my obsession, I found myself dreaming about jumping across ledges and fighting skeletal swordsmen. Yet something about playing a video game feels like a waste of my time. It's hard to see why this would be the case, since I rarely feel such wastefulness when I spend time reading a novel. It may just be a matter of the stigma that society has attached to video games: playing Zelda is bad, but reading Zola is good.

It is worth stepping back and exploring whether video games really qualify as games in the classic sense, particularly when they are played by a single player. Part puzzle, part manual-dexterity challenges, the solo-player game is closer to solitary games of patience. One might also explain them as pitting you against the code rather than another human player. Rama and I had played together as a team, but game play always involved one of us taking the keyboard while the other looked on, cheering at their side.

In 1991 I was blown away by the graphics, which seem incredibly simple by today's standards. Jordan Mechner, who created the game, was not an animator, and his first attempts to create the prince running through the dungeons were a disaster. So he filmed his younger brother on Super 8 video running and jumping in white clothes and then painstakingly traced over each frame of the film to transfer the footage to the computer. The result was worth it. "The moment I finally saw the character running across the screen, I got chills," he said. "As rough and pixel-y as it was, I recognized my brother's way of running, his physical personality. It was the illusion of life."

As for the question of whether you are playing against yourself or the computer, you might think of the computer as performing a rather different set of functions, controlling the rules and ensuring you are following them correctly. This serves to take away one conventional component of playing a game. When you sit around the board playing Ticket to Ride, you act like the algorithm implementing the code that says that playing certain trains on the board scores *x* number of points. Many of the rules of a board game are like lines of code. If this, then do that. That's one reason why computers are so perfectly suited to the gaming environment: they are experts at following the rules.

When humans have to execute the rules, inadvertent mistakes and misunderstandings can distort and even ruin the game. When we first played the collaborative game Pandemic, we totally misunderstood one of the rules. I didn't understand why the game was so easy to win. It was only when I played an online version that I understood that I had been applying the rules incorrectly. It was like I'd got a bug in my code, which meant we were playing an inferior game.

In contrast, the computer is set up to apply the rules correctly. It has been debugged and can do no wrong. Yet, at least for me, this takes away some of the charm and fun of playing a board game. I like the analog nature of the game—card not code—and I believe that as humans, one of the pleasures of sitting around the table playing a game lies in implementing the rules to determine the sequence of play.

59: Spacewar!

THOUGH MANY PEOPLE CREDIT PONG, the computer version of table tennis, with being the first computer game, that distinction actually belongs to Spacewar! Developed in 1962 by Stephen Russell, the game was implemented on the newly installed Digital Equipment Corporation (DEC) PDP-1 microcomputer in the Electrical Engineering Lab at the Massachusetts Institute of Technology (MIT). Though called a "microcomputer," the machine was actually about the size of a large fridge. Russell and his colleagues decided they wanted to come up with a project that would demonstrate the computer's capabilities, including showing off the display screen.

Spacewar! was the result. It's a two-player game in which each player controls a spaceship, the "needle" and the "wedge," in a dogfight around a star. A shot of fuel will accelerate you across the screen. The gravity well caused by the star can also be used by the ships to slingshot around the space.

One of my favorite features of the game is the shape of the universe it creates. The universe is finite, consisting of the finite computer screen. But this finite universe has no walls. If you travel off the top of the screen, you reappear at the bottom. Head off to the left, and you reappear at the right. To understand the shape of this universe, first connect the top and bottom of the screen to make a cylinder. Now join the left and right to create a bagel shape, or a *torus*. This is the shape of the universe in the game of Spacewar!—a finite universe with no boundaries. It has been proposed that our own universe is actually shaped like a three-dimensional version of Spacewar! According to this argument there exists a third direction that loops us back to our starting position, creating a universe that sits on the surface of a four-dimensional bagel, or *torus*. The mathematics of how this actually works is fascinating in its own right.

Traveling Around a Torus

If you are stuck on a surface, how can you know its shape without leaving the surface to look back down and see the whole thing from afar? For example, how did we know that our planet didn't

have a hole in the middle before we flew into space and saw it from a distance? One way to tell what shape a surface has is to make journeys around the space and explore the different properties of these journeys.

The game of Spacewar! features two distinct journeys in a straight line that you can make that never cross each other. Starting in the center of the screen, if you head left in a straight line, you will eventually reappear on the right of the screen and return to your starting position. Alternately, if you head to the top of the screen, this journey will bring you back to the starting position via the bottom of the screen, but it never crosses the first journey you made. This proves you must have a hole in your shape.

These are the two journeys you can make on a torus: going around the outside the torus and also going through the hole in the middle and back. There are no such journeys on a sphere. Straight lines on the surface of a sphere are great arcs, like the lines of longitude, and two of these always cross twice at polar opposites.

When Russell first came up with the idea for Spacewar!, it sounded great in theory, but he was intimidated by the challenge of executing its code. Describing the trajectory of spaceships would require a whole load of trigonometric functions, which sounded like far too much work. But his colleagues weren't going to let him off so easily. One of them contacted DEC and discovered they'd already written code to implement these mathematical functions. A trip up to the DEC offices secured the files, which were dumped on Russell's desk. He had no excuses now.

The power of code to simulate gravity and give players the experience of what flying a spaceship near a dense star might be like is one of the reasons that computers became such a powerful tool in making games. Scenarios that were only possible in the imagination before could now be represented with rules and game play, allowing you to see a system in action on the screen. A lot of successful video games are based on ideas of simulating physics—the physics is often completely wrong, and yet intuitively the game feels right. It's as if the virtual world of the video game has its own laws.

Because computers in these early days were so large, expensive, and inaccessible, early games tended to be two-player games like Spacewar! and Pong. Only in the late 1970s, with the advent of personal computers, did games begin to be developed for the lone player stuck in their bedroom. Multiplayer computer games started to reappear in the 1990s with the arrival of the internet, which could connect these computers.

The computer was also a perfect platform to implement the role-playing games that were taking the gaming world by storm in the 1980s. Developers started to explore how to create environments that might allow a player to alter the course of the game according to their game play. The 2010 PS3 game Heavy Rain allows players to make choices at points during the game that result in a seamless film noir experience where the narrative changes from player to player. Because your choices in the game might mean that you are responsible for the death of a central character in the closing sequences, you feel more emotionally involved than if everyone experienced the same ending.

The 2008 game Fable 2 created for Xbox sees your character morphing, becoming more or less evil, purer or more corrupt, fatter or thinner, according to actions you make as a game player. No Man's Sky, released in 2016, uses algorithms to generate a potentially endless supply of planets populated by their very own flora and fauna. Each environment is unique, created by the code when a player first visits the planet. Even the creators of the game don't know what the algorithm will produce before a player lands on a new planet.

With the advent of machine learning, where code can change, mutate, and update itself according to your interactions with it, there will be ever more opportunities for bespoke games for individual players. Games can tune themselves to the abilities of the players, so that weaker players don't get frustrated by the difficulty of challenges. The music that accompanies you as you make your way through each level is an increasingly important feature of the experience. Machine learning that composes on the fly so that the music matches your decisions and actions in the game will allow for the creation of individual soundtracks for every player, an impossibility for a single human composer. The 2021 film *Free Guy* already hints at the potential of machine learning to create an exciting array of nonplayer characters in games, although we're probably still a long way from the film's vision of one of these characters achieving free will.

60: Tetris

THOUGH MY JOURNEY AROUND the world bypassed Russia, I shouldn't leave you with the impression that Russia is not interested in games. During my time in Russia visiting for mathematical conferences, chess was frequently the game of choice, especially among my fellow mathematicians. But a video game created in 1984 in the Soviet Union took the world by storm when it crossed the Iron Curtain: Tetris.

Alexey Pajitnov had been employed by the computer center at the Soviet Academy of Sciences to work on speech recognition, but his real passion was for games. In his words, "Games allow people to get to know each other better and act as revealers of things you might not normally notice, such as their way of thinking." He particularly liked a set of puzzles that had been introduced in 1907 called *pentominos*. These consisted of a set of twelve different shapes made up of five squares, and the challenge was to tile different-shaped regions using the pentomino pieces. This was the spark for Pajitnov's idea. After some experimenting, he found shapes made up of five squares were a little too complicated, but cutting the shapes down to four squares proved a masterstroke.

Pentominos became tetrominos, *tetra* being Greek for "four." Combining the word with his other favorite game, tennis, Pajitnov called his creation Tetris.

The game is so ubiquitous that it probably doesn't need to be described. However, here goes: On your screen is an empty grid of squares, ten squares wide and twenty squares high. When the game starts, the shapes Pajitnov called *tetrominos* start descending one by one from the top of the screen. There are seven different shapes consisting of four squares:

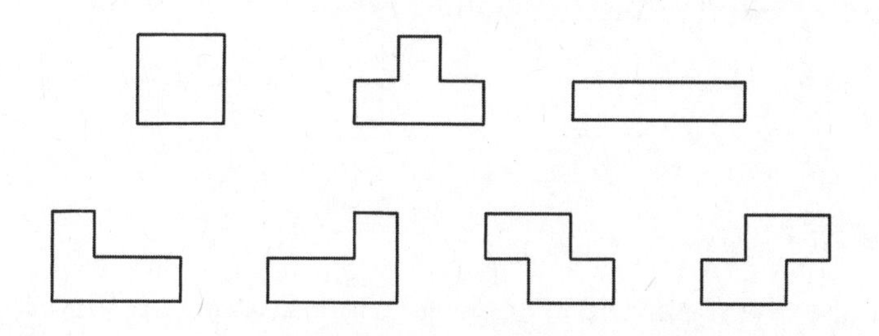

These are all the shapes you can make with four squares. As the shapes descend, you get the chance to rotate them and move them left or right. The shapes come in random order, and you get a sneak preview of the next shape as you are trying to arrange the current shape, which can help you plan a step ahead. The challenge of the game is to stack the shapes so that all the squares along a row are covered. Once this happens, this row is removed. You fail the game if the pieces pile up in such a way that rows can't be cleared, and eventually there is no room for a new piece to make it onto the board.

If you are good enough, is it possible for this game to go on forever? Heidi Burgiel, a mathematician at the Geometry Center in Minneapolis, showed that a certain sequence of tetrominos would, despite a player's best efforts, result in their losing the game. Her theorem proves that if the player encounters alternating Z pieces, then whatever they do to rotate and stack them, the game will be lost after placement of at most 69,600 pieces. The Z pieces are the awkward ones to place because they inevitably leave holes in your rows as you try to place them, and these gaps start to build up across the screen.

But isn't that a highly unlikely scenario? Sure, but at this point it is worth introducing the infinite monkey theorem. If you have a monkey hitting keys at random on a typewriter for an infinite amount of time, then, with probability of one, the monkey will at some point type out the famous line from Hamlet, "To be or not to be." It's the same principle behind why pi contains your bank account details at some point in its decimal expansion. Given that the tetrominos are being chosen randomly, the same applies to Tetris. With probability of one, at some point the machine will spit out 69,600 alternating Z pieces, just like the monkey will type out the Hamlet quote.

A Tetris Puzzle

If I take the T-shaped pieces, is it possible to arrange them so they fill a 10 × 10 grid?

The answer is no. Suppose the grid is covered in T pieces. Color the grid like a chess board. If the T pieces are to sit on the board, then they too must be colored in one of two ways: one white surrounded by three blacks, or the other way around. Since there are two more squares of one color in each T, this means that we must have an equal number of T pieces of each type. That gives an even number of pieces. But how many pieces do we need in total to cover the grid given that there are 10 × 10 squares? There are four squares in each T, so that means you need $\frac{100}{4}$ pieces. But that's 25, an odd number. We've just proved, using our coloring strategy, that there must be an even number of pieces. This implies it is impossible to cover the 10 × 10 grid using the T shapes from Tetris. Notice that the odd-even trick doesn't work on an 8 × 8 chess board, and in fact you can cover an 8 × 8 chess board with T pieces.

When Pajitnov showed colleagues his first version of the game, they became totally addicted. In a few weeks, every institute in Moscow with a computer had installed the game. People were wasting so much time playing that some speculated the game was planted by the United States to destroy Soviet productivity. The Medical Institute in Moscow had to ban the game in order to get employees back to work.

Copies of the game started to spread beyond Moscow like a virus, eventually arriving via floppy disk in Hungary in 1986. There Robert Stein, a software salesman from the United Kingdom, saw the game. He immediately saw its potential and sent a fax to Pajitnov to see if he was interested in giving him a license to sell the game. Pajitnov replied that he was open to the idea. Stein saw this as a green light and started spreading the game to the West. So began the classic saga of how others made money out of Pajitnov's invention, while he earned nothing from the game's huge success.

After a series of legal battles Pajitnov eventually wrestled back the rights to the game in 1996 and founded the Tetris Company with a Dutch entrepreneur Henk Rogers, who'd wooed Pajitnov over a game of Go. Still, some years later, he very generously said, "The fact that so many people enjoy my game is enough for me."

The game has proved so successful that it is regarded as the best-selling video game of all time. The mobile app version of the game alone has been paid for and downloaded over 425 million times. The game is so addictive that some have even risked all to keep playing the game. In 2001, a UK man was jailed for four months for playing Tetris on his mobile phone during a flight from Egypt to the United Kingdom. It should be pointed out that he was the pilot!

But what is the effect on our minds of playing computer games like Tetris for hours on end? Given the number of hours people dedicate to clearing levels and fighting battles online, many parents worry about the damaging effect that playing computer games can have on children, and there is very real research about how playing computer games can change the way we think and even restructure the brain. Take the Tetris effect, for example.

This concept describes the effects that dedicating hours to playing Tetris might have on a player. Players reported being plagued by after-images of the game for days afterward and found themselves thinking about ways different shapes in the real world can fit together, such as the boxes on a supermarket shelf or the buildings on a street. And as they fell asleep, many reported being tormented by images of Tetris blocks descending in their line of vision.

But as far as the structural effects of extended exposure to Tetris, the results appear to show surprisingly positive effects. Research done in fMRI scanners at the University of New Mexico in Albuquerque found that Tetris actually boosts general cognitive functions such as "critical thinking, reasoning, language and processing" and increases cerebral cortex thickness. A team at Oxford University has found that playing Tetris after viewing traumatic material in the laboratory reduced the number of flashbacks to those scenes in the following week. At Plymouth University researchers revealed that Tetris could distract from cravings and give a "quick and manageable" fix for people struggling to stick to diets or quit smoking or drinking.

As much as we may fixate on unwinnable Tetris games as we lie awake in bed, it appears that Tetris is good for you after all.

61: The Game of Life

AT THEIR BEST, games act like abstracted versions of the mess of everyday life. They can help us understand how a set of rules can give rise to different endgames. Among mathematicians, one particular favorite game illustrates beautifully how life itself might have emerged from simple rules of cellular growth. However, the Game of Life isn't really a game at all. Nevertheless, this program, created by mathematician John Conway, offers a perfect illustration of one of the core dynamics of great games, as well as life itself: the way that simple rules can give rise to extraordinary complexity.

I first encountered the game when I went to visit Conway in Cambridge in 1986. I was coming to the end of my undergraduate studies and had set my heart on doing research into the mathematics of symmetry. My tutor in Oxford told me how a group in Cambridge headed by Conway had just completed the most amazing piece of work documenting the building blocks of symmetry. Published in a wonderful volume titled *The Atlas of Finite Simple Groups*, this book is like a periodic table for the world of symmetry. It is still one of the most prized books on my shelf.

The project was the culmination of a 150-year journey by generations of mathematicians that started with the creation of a new language for symmetry by French mathematician Evariste Galois. Killed in a duel over love and politics at the tender age of twenty, Galois had created a new way to understand symmetry that revealed that there were certain atomic symmetries from which all other symmetrical objects could be built. These included things like prime-sided polygons like the pentagon. But as Conway and his group documented in the *Atlas*, there were also extraordinarily exotic beasts like the Monster, a snowflake that lives in 196,883-dimensional space.

Keen to play a part in the next chapter that would follow the publication of the *Atlas*, I headed to Cambridge to talk to Conway about starting a PhD in the mathematics of symmetry. But upon arriving in the common room in Cambridge, I was met by students huddled around a 19 × 19 grid placing white stones on the board. I recognized the pieces: they were from the ancient game of Go. But with only white stones on the board the students clearly weren't playing Go.

It turned out they were playing Conway's Game of Life. As I said this isn't really a game at all; rather it's an example of something called a *cellular automaton*. The idea is that you place white stones down on the squares of the board. These are regarded as live cells. A square with no stone is regarded as dead. But then certain rules kick in to determine in the next generation which cells stay live, which cells die, and which cells come to life. The fate of a square is determined by the eight squares surrounding it.

Here are the "rules of the game":

Survival: If a cell is live, then it survives to the next generation if it has two or three live neighbors in the eight cells surrounding it.

Death: A cell dies if there is overcrowding—that is, if there are four or more cells surrounding it; or it dies from isolation if its neighbors include only one or no live cells.

Birth: A new cell is born if the square is bordered by exactly three live cells.

The fun of playing this game comes from concocting new arrangements of stones on the board that give rise to interesting behaviors as the game proceeds. One important point: the game is not confined to the 19 × 19 board used for Go but is in fact meant to be played on an infinite two-dimensional grid. The game of Go simply provided a helpful practice arena for the students to start exploring the game. If you swap your Go stones for black and white pixels and a computer, you can much more fully explore the menagerie of life this game spawns.

Due to the way that the rules are written, some configurations never change at all. For example, a block of four live cells doesn't change from one generation to the next. Each live cell is surrounded by three other live cells and so survives to the next generation, but no other cell comes to life. This block of four cells is an example of something called "Still Life" in the Game of Life.

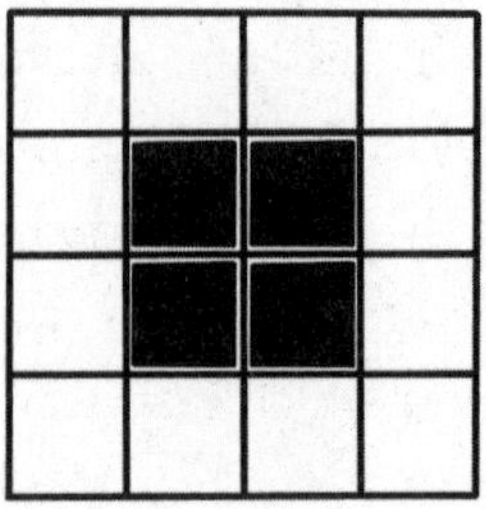

There are other configurations that oscillate. Put two blocks next to each other, and the corner blocks that touch die from overcrowding but are subsequently born again.

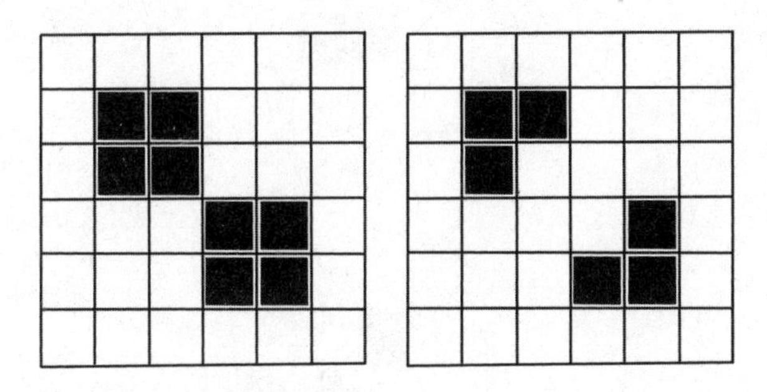

Some patterns oscillate while moving across the screen. For example, the following configuration of five cells is known as "the glider." It goes through four different patterns, but when it returns to the first one, the pattern has moved diagonally across the screen.

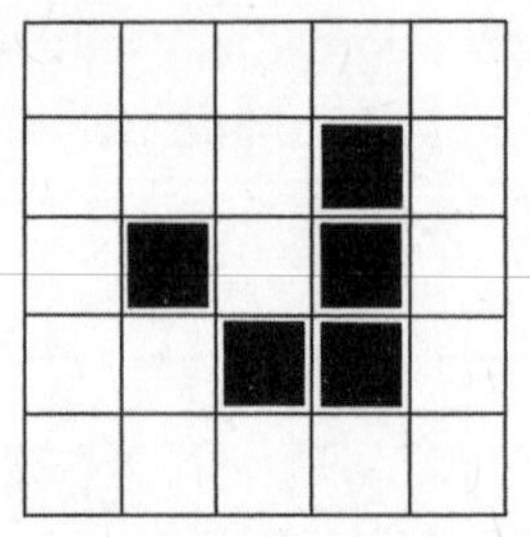

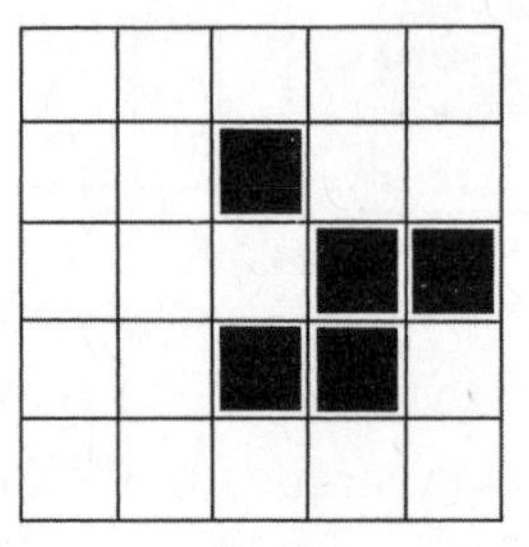

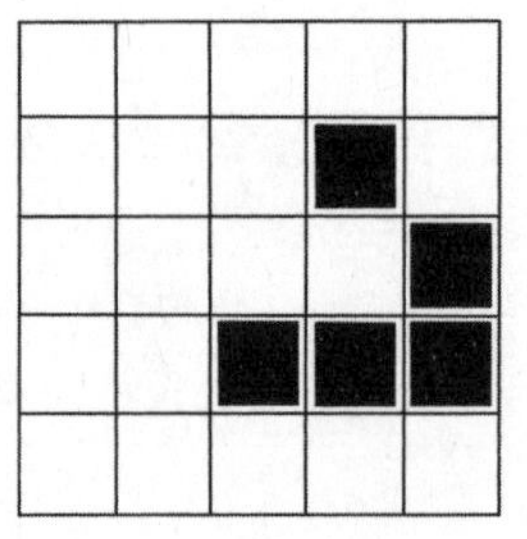

The glider is an example of a life form that will require an infinite screen to maintain its existence.

Another arrangement of five live cells looks very innocent but takes 1,103 generations before it settles down into life forms that are still, oscillate, or act like spaceships, like the glider. After Conway sent his Game of Life to Martin Gardner's *Mathematical Games* column in *Scientific American*, Gardner named these strange life forms that took so long to reveal their true nature "Methuselahs," after the biblical figure with the longest lifespan in the Bible.

Conway was interested to discover whether there were life forms that could grow arbitrarily large. His instinct was that there weren't. When Gardner published about the Game of Life in the October 1970 edition of *Scientific American*, Conway took advantage of the opportunity to offer a £50 prize to anyone who could prove him wrong and find a population of cells that could grow indefinitely. He set a deadline of the end of 1970 to claim the prize.

Sure enough, by November 1970, Bill Gosper of MIT had found a fascinating configuration that every thirty generations spat out a glider that then drifted off across the screen. Known as the "Gosper glider gun," this population, which produced more and more gliders as it evolved and never died, would therefore grow indefinitely. Gosper claimed his £50 from Conway.

But for me, one of the most extraordinary consequences of the glider gun was that, like the cards of Magic: The Gathering, it could be used to prove that the Game of Life is actually rich enough to simulate a universal Turing machine. A configuration of gliders can be set up, for example, to calculate the one thousandth prime number, as we shall explore in the math box.

One consequence of the Game of Life's ability to simulate a universal Turing machine is that the game is undecidable. This means there is no algorithm that will decide whether, given two configurations, the second pattern will evolve at some point from the first pattern.

In the Prime of Life

To use gliders to find primes, the trick is to understand that gliders, when they meet at ninety degrees, will annihilate each other. This can be used to create a *sieve of Eratosthenes*, a simple algorithm that produces a list of prime numbers. Gosper's glider gun emits a glider every thirty moves. These gliders represent

the numbers to be tested for primality. At ninety degrees to this stream of gliders, you set up glider guns that emit gliders every multiple of thirty. So the first glider gun emits gliders every 2 × 30 moves. The *n*th glider gun emits gliders every (*n* + 1) × 30 moves. The point is that the *n*th gliders can be used to remove all gliders from the original gun that are divisible by (*n* + 1). For example, if you want to discover the next prime number after seven, then find the first glider that survives passing the divisibility guns, testing divisibility from one to seven. You'll find that every glider up to the eleventh gets knocked out, but the eleventh glider makes it through. In this way the Game of Life has discovered that eleven is the next prime number after seven.

As the students in the common room in Cambridge were playing around with different configurations to see what might emerge, Conway entered the room and sighed, "I hate life." It seemed a rather harsh assessment, but he is rather embarrassed that a mere game that he sent to Martin Gardner has become one of the best-known of his creations. He always checks books' indexes to see if his name is mentioned but never reads the book if the Game of Life is also indexed. Still, he has at least admitted that he rather enjoys being introduced to give a talk as the "inventor of life."

One important point that the Game of Life illustrates is the power inherent in creating simple rules that give rise to rich and unexpected consequences. In some ways this captures not only the best games but also the best mathematics. For me it was one of the appeals of the mathematics that I was hoping to pursue for my doctorate. The rules of group theory, the mathematics of symmetry, are as simple to state as the rules for the Game of Life. And yet the *Atlas* that Conway and his team had published was an illustration of the richness and complexity that these simple rules entailed. The mathematics of symmetry is a game that I am still playing to this day.

The Game of Group Theory

Take a die and place it on the table. Consider all the ways you can pick up the die, rotate it, and then place it back down again. You will discover there are twenty-four different moves you can make, which I will call $r(1), \ldots, r(24)$.

All these moves satisfy the following rules, which make up the game of group theory that I love to play.

1. If you do one move $r(i)$ followed by another $r(j)$, then the combination must be a third move $r(k)$ that you could have done in one go. This creates an interesting algebra that connects the twenty-four moves:

$$r(i) \circ r(j) = r(k).$$

2. The moves include a strange move $r(1)$ where I don't rotate the die at all. This is like the number zero. Combine it with any other move, and you don't get anything new.

$$r(1) \circ r(i) = r(i) \circ r(1) = r(i).$$

3. Every move $r(i)$ has another move $r(j)$, called its *inverse*, that returns the die back to its original position. We can write this as

$$r(i) \circ r(j) = r(1).$$

4. Finally, if I take three moves $r(i)$, $r(j)$, $r(k)$, then I get the same result if I do two different ways of combining them:

$$(r(i) \circ r(j)) \circ r(k) = r(i) \circ (r(j) \circ r(k)).$$

This is called the *rule of associativity*.

If I have any mathematical object and a list of moves that I can make, then provided they satisfy these four rules, that object qualifies as a symmetrical object. The moves are its symmetries.

The game I play is to create new symmetrical objects that no one has discovered before. Often these objects can't be seen in our three-dimensional world. They are like uncovering the dice in hyperspace.

62: Tic-Tac-Toe or Noughts and Crosses

WE HAVE SPENT OUR INTERLUDE exploring the solitary and addictive nature of computer games, the question of whether you are playing against the computer or the computer is simply executing a game's algorithm, and the way that computers can embody the simplicity and complexity of the best mathematics. What remains is the fascinating spectacle of computers themselves joining in as players of various games. One of the first games that a computer learned to play is one that begins many kids' journeys into gaming: the classic pen-and-paper game tic-tac-toe, or as it is called in the United Kingdom, noughts and crosses. The English name derives from the name for the O that gets placed in the 3 × 3 grid in opposition to the cross X. This is a game whose history is very old, with evidence of 3 × 3 grids carved into the ground throughout sites in ancient Egypt. But it is also a game that has been continually reinvented independently across the world from Korea to India to Ghana and by the Zuni Native Americans.

Because Os and Xs don't move once played, it is a fairly easy game to analyze, and it isn't too long before kids latch on to the strategy to force a draw.

Indeed this is a game that, played perfectly, will always end in a draw. Regarding games that are rotations or reflections of each other as the same endgame, there are only 138 configurations that the game can end in. Of these ninety-one are won by X and forty-four by O; only three end in a draw. But if each contestant plays carefully, they can always guide the play into these last three states.

Given the symmetry of the board and the tree of possibilities for where the game can end up, it is quite an easy exercise for first-time coders to write instructions for a machine to mindlessly follow—it did not take long in the modern history of computing for programmers to write the code that allowed a computer to play along.

Code to play tic-tac-toe was implemented in 1952 in OXO, one of the first digital computer games with video output. It was programmed by British computer scientist Sandy Douglas for his Cambridge PhD thesis project. Cambridge was then home to the very first stored-program computer, the Electronic Delay Storage Automatic Calculator (EDSAC). The program was input via punch cards, the player made their move via a telephone dial whose numbers corresponded to positions in the grid, and then the output was displayed on the in-built oscilloscope's cathode-ray tube readout display.

The program to play the game is simple enough that one didn't really need the electronics of the EDSAC to run it. Indeed a beautiful example of the mechanical nature of playing the game was implemented by a group of MIT students who used Tinkertoy pieces from a kids' building kit to create a machine that could play the game perfectly. Yet, with such simple parts, analyzing the tree of possible games meant that the machine still had to consist of ten thousand wooden pieces held together with fishing line and about four thousand brass pins.

The machines in both these examples are just mindlessly implementing the clever analysis done by humans to understand how to play this game. This is a far cry from machine learning, which describes what happens when the code responds to a failure to achieve its target, in this case when it loses the game, by updating its code, based on what it has learned, to play the game differently the next time round. Machine learning has more recently revolutionized computers playing games like Go, but it is not quite as recent an innovation as people think—in fact it was created within a decade of the first computer programmed to play tic-tac-toe.

The first example of machine learning in game play also consisted of a machine that learned how to play tic-tac-toe. But rather than being made from electronics, this machine consisted of 304 matchboxes. Created by British computer scientist Donald Michie in 1961, it was called MENACE, which stood for Machine Educable Noughts and Crosses Engine. Each matchbox corresponded to all the possible configurations the game could be in from beginning to end. Again, you didn't need to consider every state because you used symmetry to reduce things down to 304 different configurations. So a match box might have on its side

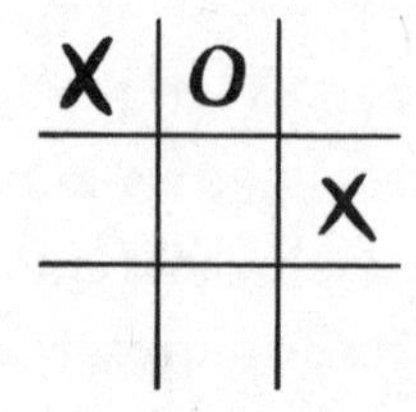

This would be the box you would go to if you saw any symmetrical variation on this arrangement—for example, the reflection

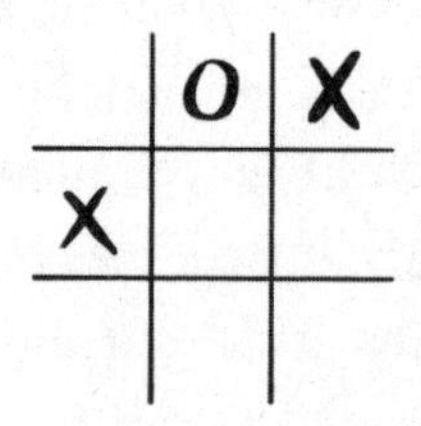

or the rotation

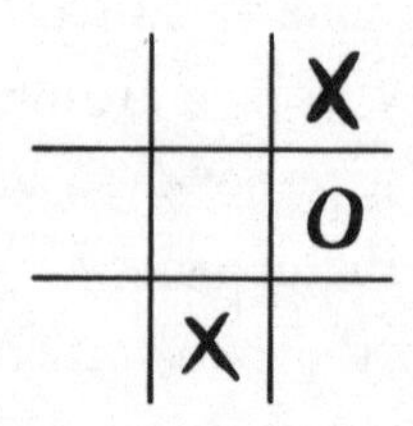

The matchboxes contained colored beads. Each color corresponded to a position in the grid.

White	Lilac	Silver
Black	Gold	Green
Amber	Red	Pink

If you were mid-game, to find out the machine's move, you would go to the matchbox corresponding to the current configuration and randomly select a bead from the box. This would tell you the next move for the machine to make. At the beginning of the learning process, the matchbox with three moves made already would have six colored beads corresponding to the next possible move.

The learning process was implemented by the way the beads were redistributed after each game played. The idea was that a matchbox that corresponded to a bad choice would gradually lose those beads while the strong moves would be rewarded by adding three more beads of the color that helped win the game. A draw was also rewarded by adding an extra bead of the color used by each box in the game. As the machine learned, it would play more optimally because the beads would tell it the better move to make each time.

If MENACE was unlucky, it might play nine losing games in a row. As a result, the first box would have been emptied of all beads, which meant it would resign before it even started. So the learning was not guaranteed. This scenario happened in about one in ten runs of the learning process.

Over a weekend Michie played the matchbox machine, redistributing beads as the machine won or lost or drew. After only fifteen games with Michie playing optimally, MENACE had already learned to abandon making its first move in noncorner positions. After two hundred games, the machine was able to cope with whatever optimal strategy he tried to implement, forcing a draw each time.

Tic-tac-toe has been the inspiration for some intriguing variations, like 3-D tic-tac-toe played on a 3 × 3 × 3 board with twenty-seven different positions. This game can always be won by a player making their first move in the center of the game. Connect 4 is another game inspired by tic-tac-toe where gravity plays a part in limiting where you can play. First published in the 1970s by the American board game manufacturer Milton Bradley, this is another game where the player going first can guarantee a win by the forty-first move (there are only forty-two moves possible since the board is 6 × 7 in layout). This was only proved in 1988 independently by James Dow Allen and Victor Allis. (If you'd like to accomplish this, here's a hint: start by putting your counter dead center on the board.) Connect 4 has shown that there is still an appetite for games that involve getting your counters in a row, having sold over four million copies by 2018.

There is even a crazy quantum variation of the game of tic-tac-toe. In quantum physics an object can be in two places at the same time before it is observed. So in quantum tic-tac-toe, you can play your X or your O in two different squares. These squares are then entangled. But because the X or O hasn't committed to a square, these are marked in pencil. As play goes on, squares become more and more entangled. The wave function collapses, for example, if Square A is entangled with Square B, Square B is entangled with Square C, and Square C is entangled with Square A. These are called *cycles of entanglement*, and they can vary in length. They cause the quantum state to collapse, forcing the Xs and Os to commit to which square they are occupying. The quantum becomes classical, and the Xs and Os get penned in.

From the temples of ancient Egypt to the era of machine learning, it's remarkable how such a simple game has provided three thousand years' worth of entertainment and education for the human species. Yet I'm not exactly sure whether the advent of computers playing games has been a good thing for the history of games. Once you know that there is a computer out there that can play chess better than any human, doesn't that take the edge off playing the game? Isn't it now just a game for us dumb humans to play against each other?

I tend to get the same feeling whenever I discover that a game has been "solved." In other words, mathematicians or computers have found the optimal strategy to play Connect 4 or checkers. Again, it makes one feel that the game is only really interesting when it's played by people who aren't playing optimally. This is why I maintain that we always need an element of chance involved in our games—in order to give the suboptimal player a chance!

Even here it seems that computers maintain an edge. Poker is one-third mathematics, one-third human psychology, and one-third chance. And yet computers playing poker have succeeded in taking the chips when they come to the table against human competitors. It seems like computers have not only mastered math and human psychology but also minimized the impact of chance in the game.

Of course, one of the exciting benefits of the involvement of the computer is to teach us humans new ways to play our games. Chess grandmaster Magnus Carlsen says that he plays a much more aggressive style of chess after learning how computers play the game. And Go has gone through a third renaissance in style of play thanks to the discoveries made by AlphaGo and its descendants.

One of my favorite AI movies perhaps illustrates what we have in store. In the film *Her*, following his divorce the central protagonist purchases an AI assistant called Samantha. He ends up falling in love with the assistant but is horrified to learn that the operating system is maintaining relationships simultaneously with thousands of other humans. Ultimately, the AI just finds interactions with humans too slow and dumps all the humans to have a relationship with another AI.

There is something to this. As AI evolves, it will seek not just relationships with other AIs. In time the art AI produces, the music it composes, and the novels it writes will probably be of a complexity that only another AI will appreciate. The same goes for the games AI will play. Humans will be left behind to play among ourselves while the AI enjoys a game with another AI. But perhaps we have already accepted this state of affairs. After all, the one-hundred-meter race is one in which only humans compete. No one is expecting to beat a sports car or a jet engine to the finish line.

Our finish line is growing closer, though there are still a few more stops to go. Phileas Fogg made his way across the Atlantic straight to England, but we are going to take another detour to explore some of the games that have emerged from Africa.

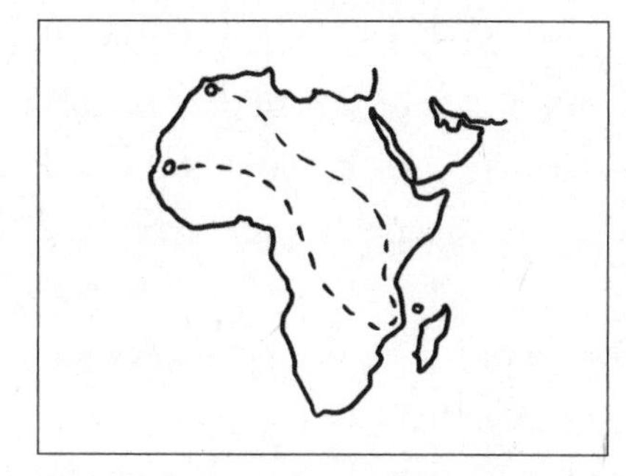

CHAPTER 14

AFRICA

ALTHOUGH I'VE SPENT TIME IN northern Africa, I have never traveled further south than the Sahara Desert. That is a shame, because there are many countries in Africa on my bucket list that I want to visit: for the music, culture, landscape, animal life, and much else. But for now I shall have to content myself with the alternative way of exploring the continent that our journey in games provides. It is in some sense one of the intents of this book, and perhaps a role of games in general, to whisk you off for an afternoon to far distant lands that you can't visit.

I have often asked friends and colleagues on trips abroad to keep a keen eye out for interesting games. One game in particular is encountered by nearly all those who have traveled in Africa, regardless of country.

63: Mancala

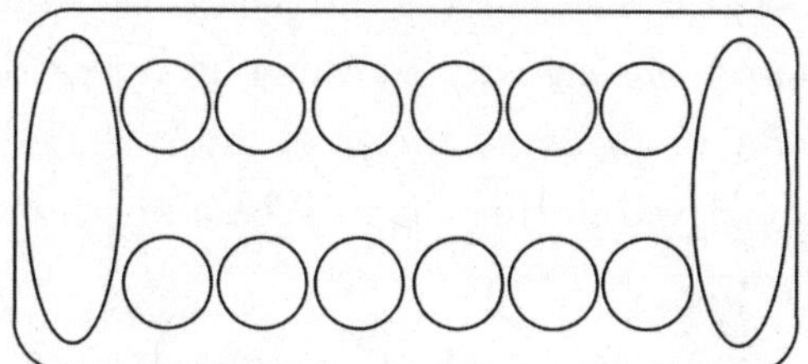

MANCALA IS ONE OF THE oldest two-player strategy games that we'll encounter on our journey round the world. There is evidence of this game being played across Africa and Asia as many as eight thousand years ago. It ranks alongside chess and Go as one of the great games of pure strategy.

The classic board consists of two rows of pits with a single pit at each end, although there are more complex boards that can have four rows. These boards have been found carved into pieces of wood, pottery, and even the roofs of houses. The designs range from simple carved pieces to hugely elaborate constructions depicting canoes, crocodiles, fish, and even a model of the local village. This variety encompasses the diversity of ways the game itself might be played, since mancala is actually just a generic name for a whole host of different games that use the board, though all share a similar character. Seeds or stones are placed in the pits, and then players take turns picking up the seeds in one pit and redistributing them one at a time in the neighboring pits. There are rules for the capture of seeds in a pit that are then collected in the two end pits. If your last seed lands in an empty pit, you get to take all the seeds in the pit opposite. No luck is involved; this is a game devoted to strategy.

In Africa the game is not just popular for the fun of playing; it also has spiritual importance. Ancient cultures, including those in Africa, recognized forty-eight different constellations in the night sky, often associating them with gods. This could be why forty-eight seeds will generally be used in the game. Just as the pieces in the Royal Game of Ur symbolize the movement of the planets, perhaps mancala is a representation of the forty-eight constellations in the night sky. Or it could simply be that forty-eight seeds divide nicely: with six pits on each side of the board, that allows four seeds in each pit. Additionally, the board is laid out on an East-West axis to honor the souls of the dead ancestors who are believed to live in the West. In western Africa the game is often played during the dry season because it is meant to encourage the rains.

A number of rules and regulations surrounding the game serve to acknowledge its power. In Mali many villages ban the playing of the game by children. In the Ivory Coast it is believed that a pregnant woman who plays the game can influence the sex of her unborn child by playing an opponent of the sex she is hoping for.

Once the sun sets, people are much more reluctant to play the game as it has the power to attract malicious spirits. Sometimes the boards are left outside for the spirits to play. The Baule people of the Ivory Coast make a point of exploiting the potency of the game at night to help decide who will be a new leader of the village. A new king of the people of Ganda would be required to play and win a game of mancala before he could assume power. Though different cultures played slightly different variations of mancala and regulated the game in their own ways, the game fundamentally served as a powerful bond across groups that might not share a common language. The game's universal nature allows it to transcend cultural and geographical boundaries.

The oldest known mancala board was found not in Africa but in Jordan during excavations in 1989 and dates back to nearly 6000 BCE. Two lines of six holes are carved into the limestone floor of a Neolithic house in 'Ain Ghazal. In which direction the game crossed the Red Sea is unclear, but it certainly took hold as one of Africa's most popular games, and the continent is home to many of the complex variants that have evolved.

A whole range of different versions of the rules results in an increase in game complexity. The Yao ethnic group in Malawi enjoy Bao, a complex version of mancala with four rows of pits; the Buganda in Uganda tend to opt for Omweso, a version also played on this larger board in which you can resow captured seeds back into the game. In contrast, the Ashanti in Ghana and the Mende in Sierra Leone prefer the simpler two-row board with the simple set of rules that most players start playing with. Does a country that enjoys tougher versions of the game have more innovative business communities? There is some evidence of a connection, but whether playing games is the catalyst for business success is hard to establish.

There have been some fascinating suggestions that mancala might be favored over a game like chess in Africa due to the egalitarian nature of the game. Each seed has equal value.

Seeds are not destroyed but redistributed according to the player's abilities. In chess, the pieces represent a highly hierarchical society from the king and queen down to the pawns. The game is destructive. But whereas chess is war, mancala is the market. Along these lines, there has been interesting research into whether the playing of the game correlates with the development of a more entrepreneurial outlook across different groups in Africa. Mancala requires planning ahead, not leaving your seeds vulnerable to capture, and understanding the consequences of each possible move.

Despite its simple nature, some versions of mancala are mathematically so complex that we still have not understood the optimal game play to put yourself in the driving seat. This is particularly the case in versions of the game where you are allowed to resow captured seeds. This reseeding leads to a game whose complexity does not relent as the game proceeds. But in versions where seeds, once captured, remain out of play, the game gets simpler as more moves are made, and we gradually see a convergence to a winning side. In these simpler versions, a brute-force analysis of the possibilities can point us to winning strategies or at least strategies that might force a draw.

The following table provides an analysis of what you can expect if the game is played perfectly in different scenarios where the number of holes and seeds can vary. Columns denote different numbers of seeds in the pits. The rows represent the number of pits on each side of the board. A "W" indicates that the first player can force a win. A "D" indicates that the players can force a draw. An "L" indicates that the first player loses in perfect game play.

	1	2	3	4	5	6
1	D	L	W	L	W	D
2	W	L	L	L	W	W
3	D	W	W	W	W	L
4	W	W	W	W	W	D
5	D	D	W	W	W	W
6	W	W	W	W	W	W

For the classic six-hole game with four seeds, this analysis reveals that if you start, you can always force a win.

The Perfect Play

Let's number the holes one to six (from left to right). Mancala is generally played in an counterclockwise direction, and its rules determine that you go again if the last seed lands in your home area. With those basics in mind, here is the sequence that you need to play, together with the optimal sequence that your opponent should play to minimize the margin of the win:

36-21-4-4-6264-2-5-6-156-5-646563-64-5-2-4-3-1-65-2-4.

A dash indicates that play changes hands. Play in this manner, and you'll win by ten seeds.

Notice that in the winning sequence, there is a set of moves that allows you to play six times in a row due to clever seeding. On a six-pit board the longest sequence of moves possible is seventeen in a row, which can be achieved when clearing the following setup of stones.

See if you can clear all the seeds on this sample board.

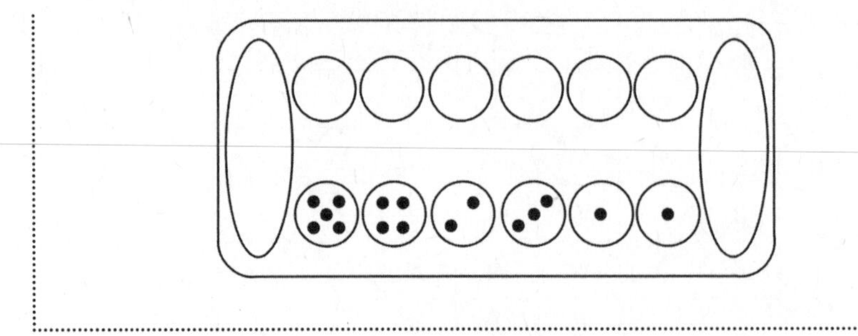

Though mathematics often gives rise to interesting new games or clever strategies for playing, given the intimate connection between the two fields, sometimes games themselves can stimulate new mathematical breakthroughs. There is evidence that mancala might have been the inspiration for one of the first accurate estimates of pi.

When I went to Egypt a few years ago, I picked up a beautifully carved mancala board in the souk in Cairo. It didn't come with counters, so I visited the spice shop next door and bought a bag of some unidentified spherical seeds about the size of marbles that seemed a good size to play the game. It was seeing the seeds sitting in the circular pits that apparently inspired an idle game player's discovery of an estimate for pi.

Pi defines the ratio of the circumference of a circle to its diameter. Whatever the size of the circle, this ratio will always be the same. If a circle is one meter across, then the number of meters it takes to go around the outside of the circle is 3.14159...and then the numbers spiral off to infinity in a dizzying dance of digits.

Calculating an exact value of this important number has been a consistent obsession throughout the history of mathematics right back to ancient times. There is even a calculation of pi in the Bible. 1 Kings 7:23 reads, "Ten cubits from one brim to the other, it was round all about and a line of thirty cubits did compass it about." This gives a value of pi as exactly three, although if this were its true value, circles would actually look more like hexagons.

One of the very first significant mathematical documents in the history of my subject is the Egyptian Rhind Mathematical Papyrus, written by the Egyptian scribe Ahmes in about 1650 BCE. Housed in the British Museum, it is full of fantastic mathematics, including ideas for how to use binary numbers to do multiplication three thousand years before German mathematician Gottfried Leibniz would reveal their potential. There are cunning discussions of fractions and puzzles about geometric series. And significantly for the architects of the time, there is also the first calculation of the volume of a pyramid, proved using an early forerunner of calculus.

But contained in this papyrus is a calculation that leads to one of the first estimates of a value for pi. Ahmes tries to estimate the area of a circular field whose diameter is nine units across. Because the area of a circle is pi times the radius squared, if he knows the area and the radius then he can calculate pi. Ahmes's calculation of the area is particularly striking because it depends on seeing how the shape of the circle can be approximated by other shapes that the Egyptians already understood.

The Rhind papyrus states that a circular field with a diameter of nine units is equal in area to a square with sides of eight—but how would this relationship have been discovered? My favorite theory sees the answer in the ancient game of mancala.

How to Use Mancala to Estimate Pi

As the players sat around waiting to make their next move, perhaps one of them meditated on how pleasingly the stones filled the circular holes in the mancala board. For example, you can have one stone in the center surrounded by six stones in a hexagonal shape around the central stone.

The player might have gone on to experiment with making larger circles and discovered that sixty-four stones can be arranged to approximate a large circle with a diameter of nine stones. But sixty-four stones can also be rearranged into an 8 × 8 square.

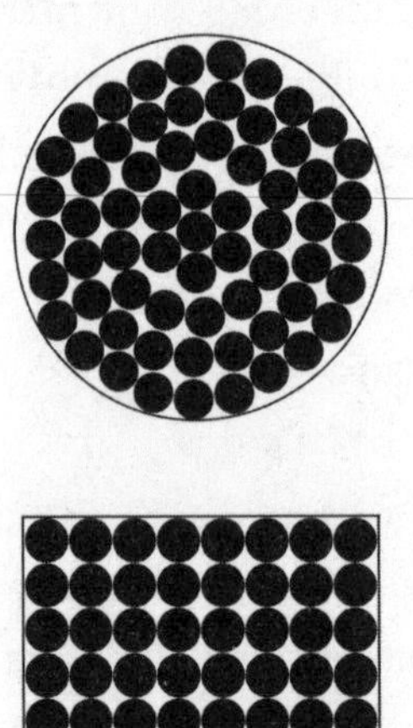

So now our mancala player has provided a method by which Ahmes can estimate the area of the circle whose diameter is nine units. He proceeds by arranging the stones that the circle has been approximated by into a square whose area is sixty-four units. Recall that the area of a circle is pi times the radius squared. The radius in this case is 4.5. So Ahmes's calculation gives the first estimate for pi as the area 64 divided by the radius squared, 4.5^2, which comes out at 256/81, or approximately 3.16. Not bad for a first estimate.

64: Gulugufe and Fanorona

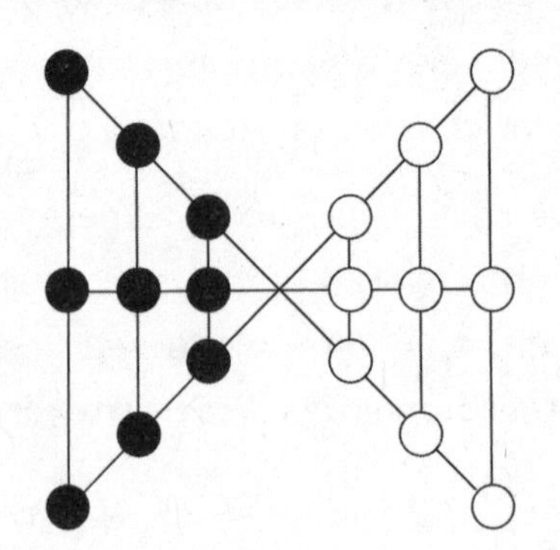

MANY OF THE OTHER GAMES that one finds across the African continent have a very similar flavor. Simple ingredients, like a playing board drawn in the earth where sticks and stones are laid down, proceed to give rise to complex strategy games. Such a game can found in Mozambique on the east coast of Africa.

Called Gulugufe, which means "butterfly" in Chechewa, the name refers to the shape of the board that the game is played on. It is basically a version of checkers or draughts, where you try to jump your pieces over your opponent's to remove them from the board. But the board is unique. Instead of a classic 8 × 8 grid, the game is played on a network of nineteen nodes arranged as two triangles of ten nodes joined at the apex of the triangles. The two triangles look like butterfly wings, which is how the game got its name. Although it is often played on boards drawn in the sand, elaborate boards can be found where the grid is laid out on an image of a butterfly replete with wings, abdomen, and antennae.

The same game can be found in India, where it is known by the name Lau Kata Kati. Given the trade that ran between India and the east coast of Africa, it seems clear that these butterfly-shaped gaming boards also traveled back and forth. In India a whole range of similar games are known as Indian war games. Some of the variants are played on boards with extra nodes attached. There are also circular variants of concentric circles joined by lines. They all have the interesting quality that the first move must see a piece occupy the central node joining the two opposing armies, which is then sacrificed by the opponent's response.

Just across the sea from Mozambique is the island of Madagascar, where you can find a rather sophisticated extension of Gulugufe called Fanorona.

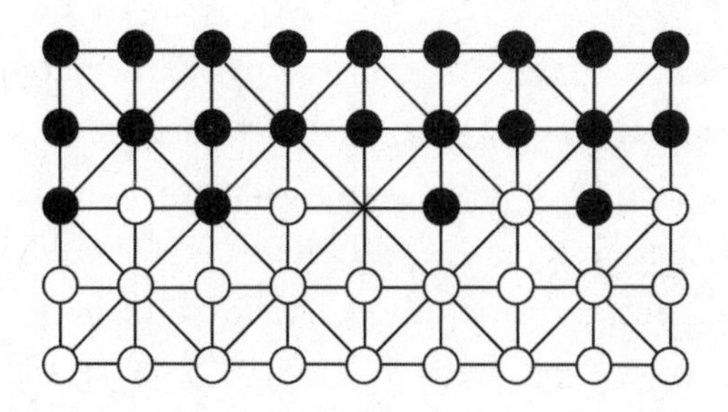

The board is larger, but the setup is similar in that the first move is into the one unoccupied point on the board. But there is a fascinating extra twist to this Madagascan game. You capture your opponent's pieces not by jumping them but by charging them or, more intriguingly, withdrawing from them. If you move a piece toward a line of your opponent's stones, they all get wiped out. Similarly, if you are next to a line of your opponent's stones and you make a move withdrawing from them, your opponent's stones are removed. And even more excitingly, in one turn you can keep combining moves to wipe out several lines of your opponent's stones. Playing the game can produce moments where your stones disappear in a combined sequence of charges and withdrawals. The other important rule that keeps the game very dynamic is that if you can capture, you must. This holds true even if the capture puts you in a vulnerable position.

The game is so compelling that it is blamed for distracting a would-be ruler of Madagascar from claiming his throne. In the early seventeenth century, King Ralambo decided to choose as his successor the first of his sons to return to his side after hearing that he had fallen ill. He was faking the illness and just wanted to test his sons' allegiance. The king's messenger, who was sent to the eldest son, arrived to find Prince Andriantompokondrindra engrossed in a game of Fanorona. The game had reached an intriguing endgame called *telo noho dimy* in which his three stones faced off against five of his opponent's stones. He wasn't going to be distracted by news of his ailing father and sent the messenger away. This delay gave his younger brother the chance to reach his father first and claim the throne. For the sake of winning a game, Prince Andriantompokondrindra lost a kingdom!

Fanorona has now been investigated by mathematicians in its entirety, and their analysis isn't pretty. After what amounts to brute-force computer inspection of all possible moves, mathematicians have shown that if you play perfectly on a 5 × 9 board, you should always be able to force a draw. But if you were to vary the dimensions of the board, things could change. On boards that are 7 × 5, perfect play will always lead to white winning. But switch the dimensions to 5 × 7, and even if everything else is the same, the advantage goes instead to black.

All of these games are forerunners of the European version of draughts (or checkers, as it is known in the United States), a game that many hold in low regard as one that kids play before they move on to chess. "The game of draughts is played by a maximum of persons with a minimum of intelligence," wrote W. G. Grace about the game. But Edgar Allen Poe is among those who gave the game far more respect: "The higher powers of the reflective intellect are more decidedly and more usefully tasked by the unostentatious game of Draughts than by the elaborate frivolity of Chess." He believed the way different pieces could move in chess was mistakenly interpreted as being more sophisticated but actually obscured the rather shallow nature of the game compared to the austerity of draughts. The game's African origins and cousins in Gulugufe and Fanorona reveal a far richer story.

65: Achi

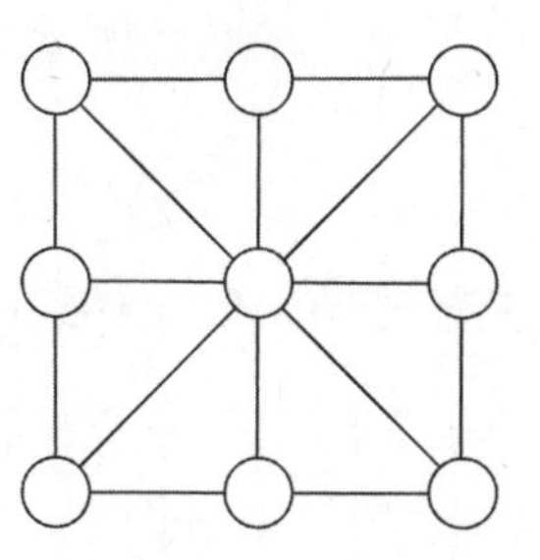

FRIENDS WHO'VE SPENT TIME in Ghana on the west coast of Africa have returned with reports of a popular game that seems to get reinvented multiple times across the globe. Called Achi, it is a dynamic version of tic-tac-toe played on the same 3 × 3 grid.

Each player begins with four stones, which they lay on the board in turn. Once all eight stones have been placed, there is one free space, which the first player can now move one of their stones to. Players take turns moving a stone of their color with the aim of getting three stones in a row. In optimal play, the winner is the person who goes first, and much as with tic-tac-toe, the winning opening move is to occupy the center spot. Yet the game is much more fun than tic-tac-toe because the dynamic nature of the game means that there will always be a winner rather than an inescapable draw.

Versions of this game appear across the world. In the Philippines it is called Tapatan and is played with three stones each rather than four. In India it is known as Tant Fant, and the three pieces start lined up at the top and bottom of the board. The Zuni Native Americans and the Pueblo people of southwestern America play a similar game, called Picaria, on a slightly more elaborate board. Pieces are placed on the intersection points of the board, giving four additional locations where stones can be placed.

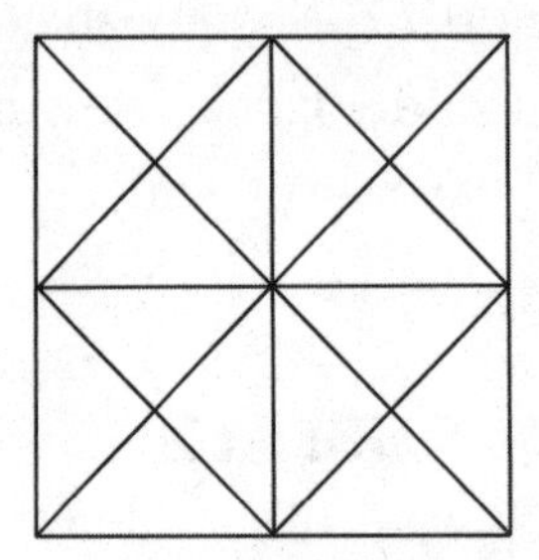

66: Bolotoudou

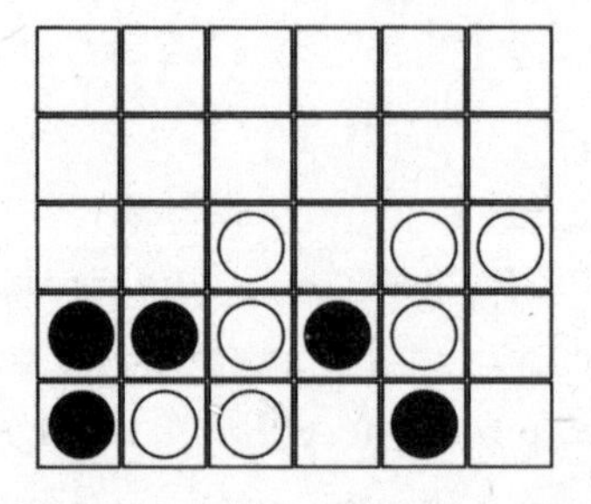

A RELATED GAME THAT IS beloved of the Sudanese is Bolotoudou. Played on a 5 × 6 board by two players with twelve pieces each, this game is interesting in allowing players to lay down two pieces at a time. Often one player has pebbles while their opponent has little sticks. The game has such cultural resonance that players will often carry their own set of pieces to play, which will be handed down through the generations.

As with Achi the aim is to move pieces to try to get three in a row. But rather than winning you the game, this instead gives you the right to remove one of your opponent's stones adjacent to the row you've made. The aim is to reduce the opponent to two stones so they can no longer make three in a row.

As with Achi, winning the game comes down to the initial layout of your stones. In Achi you can easily guarantee a win if you go first. Similarly Bolotoudou relies, more puzzle like, on finding a clever way to lay the stones to force a win in the subsequent play. Winning configurations of stones are traditionally handed down through the generations along with the set of game pieces.

Bolotoudou and Achi share strong resonances with a third game that can be found across the world. I've placed the game in Africa partly because it shares so much in common with the continent's other games but also because the earliest known board was found carved into a temple in Egypt. It seems to be a popular piece of graffiti. I last stumbled on it in Venice of all places.

67: Nine Men's Morris

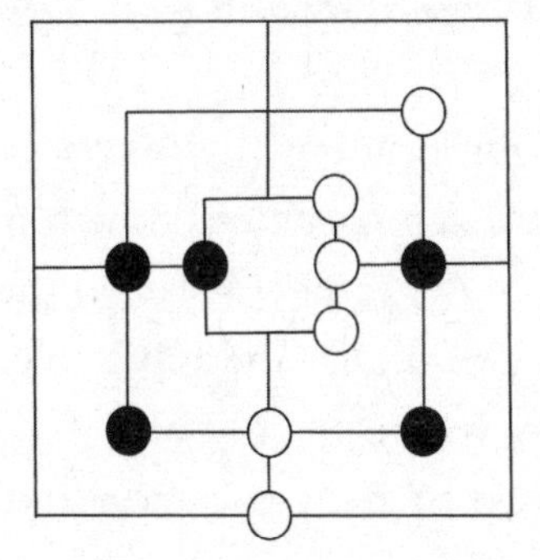

VENICE HAS A GREAT TRADITION of gaming. "Here everyone plays everything," wrote seventeenth-century adventurer Giovanni Francesco

Gemelli Careri, whose travels around the world are said to have inspired Jules Verne. Although the authorities would try to ban gambling in the city, areas reserved for gaming persisted nonetheless. Engineer Nicolò Starantonio Barattiero, who built the Rialto Bridge and managed in 1172 to raise the two great columns you encounter in Piazza San Marco, was rewarded by the doge with the license to play dice games between the two pillars when gambling elsewhere in Venice was strictly forbidden. As we have seen, Venice would go on to see the birth of the first casinos in history.

On a recent trip to the city, I found another game that the Venetians clearly liked to play. I'd just overdosed on Tintorettos in the extraordinary Scuola Grande di San Rocco and needed to catch my breath. I sat down on the stone ledge just outside the entrance to the Scuola and discovered a strange bit of graffiti carved into the stone. It consisted of three concentric squares connected by lines. I immediately recognized the image as the board used for another strategy game that, like Bolotoudou, has been around since ancient times, called nine men's morris.

The game has been played for millennia and has pretentions to African origins. A board similar to the one I stumbled on in Venice has been found carved into the roof of a temple dating back to 1400 BCE in Kurna in Egypt, and it is thought to be the oldest known example of the board. It was a game that the Romans also enjoyed, and the word *morris* is derived from the Latin word *merellus*, meaning "game piece."

Each square has eight locations that pieces can occupy, so twenty-four locations in total. Each player has nine pieces that get played in turn at the beginning of the game. Once laid they then can be moved along paths to different locations. The aim is to get three of your pieces in a line, which allows you to remove one of your opponent's pieces. As with Bolotoudou, you win if you reduce your opponent to two pieces.

Ovid refers to the game in *The Art of Love*: "There is another game divided into as many parts as there are months in the year." This is perhaps a reference to the twelve corners of the three squares. "The winner must get all the pieces in a straight line. It is a bad thing for a woman not to know how to play, for love often comes into being during play." In many cultures, games are used as a safe space for two lovers to get to know each other, and Ovid suggests that this game served the same purpose in his day.

So popular was the game that it survived well into the medieval period in Europe, with boards found carved into cloister seats in several cathedrals in England, including in Canterbury and Salisbury. Obviously the priest's sermons were not terribly interesting if the congregation felt the need to carve ad hoc boards into the pews.

The game even gets a mention in Shakespeare's *A Midsummer Night's Dream*. The fairy queen Titania at the beginning of Act 2 describes the chaos that reigns because of Oberon's actions. The countryside is in such disarray that "the nine men's morris is filled up with mud." It is possible that Shakespeare's reference is responsible for branding the game with what remains its current name. There is a large nine men's morris board marked out in the grass near the theater in Stratford-upon-Avon, where Shakespeare's plays are performed. But the board seems to have been laid down early in the twentieth century in celebration of Titania's mention of the game.

As well as being the stage for a popular game, the three concentric squares might also have a spiritual significance representing the physical, intellectual, and spiritual realms. Solomon's Temple in Jerusalem is also modeled on this design, with the twelve corners of the squares representing the twelve gates to heaven described in the book of Revelations and the four lines across the design symbolizing the four rivers of paradise. Once again we see the close link between games played purely for entertainment and games capturing a much deeper ritualistic quality.

After thousands of years of playing the game, a mathematical analysis of nine men's morris in the 1990s revealed that, as with tic-tac-toe, if you play carefully you can always force the game to be a draw. The analysis required computer brute force to explore the possible games that can be played, but because there are twenty-four locations on the board, which can either be empty or occupied by a black or white counter, an estimate of the number of states that a game can assume is 3^{24}, or roughly two hundred billion. Not all these are logistically possible in the game, but it gives an upper bound for the states that the analysis will have to consider. Once the rules of the game are taken into account, we know the pieces can be in one of 7,673,759,269 states. This is where the computer takes over to establish that there are indeed ways to play the game that ensure you can never lose. And so, after thousands of years of people enjoying playing nine men's morris, it would seem that the computer has finally succeeded in killing the game off.

Tips for Playing Nine Men's Morris

As with tic-tac-toe, the best locations to place your stones are those with the most lines intersecting that position. In tic-tac-toe, that's the center square. In nine men's morris the best places to occupy when laying your opening stones are the four intersection points on the middle square, each with four lines emerging from that point. That will give your piece lots of options for moving. The next most optimal are the points on the sides of the outer and inner squares. The weakest points, unlike in tic-tac-toe, are the corners since they have only two directions you can move your piece to.

68: Agram

MANY OF THESE AFRICAN GAMES are quite similar in spirit. The boards might vary, but they are often about laying stones on the board, moving them around into positions of alignment, and removing your opponent's stones from play. Most of these games require little equipment beyond a few stones or sticks and some sand or earth to draw the board on the ground.

But not all games are made from such simple ingredients. Card games can also be found across the continent. Agram, for example, is a card game that is very popular in Niger. It is a trick-taking game played with a deck of thirty-five cards. The court cards and, weirdly, the threes are removed from a conventional deck along with the ace of spades (known as the chief card). Each player gets six cards. The game is rather unusual in that the winner of each round is the person who takes the last trick. In Ghana the game is known as Spar or Sipa, and in Cameroon it is called FapFap. It is believed that these games probably had their origins in Europe and came to Africa via the slave trade.

But games traveled both ways. The slave trade is probably responsible for mancala finding its way to the shores of America. Games were a way for those whose freedom had been taken away from them to assert some agency. They were also a way to establish some social hierarchy among such groups. You didn't need to own possessions to compete at a game like mancala, which could be played in pits dug into the ground with stones collected out in the field.

Games have also been used as a way to communicate the horrors of what is called the Middle Passage, the transport of people from Africa to be sold into slavery in America. Brenda Romero's game New World evolved out of her attempt to teach her seven-year-old daughter about this tragic period of history. From talking to her daughter after her class had been taught about the slave trade at school, it became clear that she hadn't really grasped the horror of what these people went through. "This is an incredibly significant event, and she's treating it like, basically some black people went on a cruise; this is more or less how it sounds to her."

So Romero decided to make a game to help her daughter understand the reality of what those on the boats experienced. By the end of the game, her daughter was in tears as she recognized that many didn't survive the crossing and that, even when they did, many families were ripped apart. It revealed how a game need not just be about entertainment and distraction but can be a powerful vehicle for education and change.

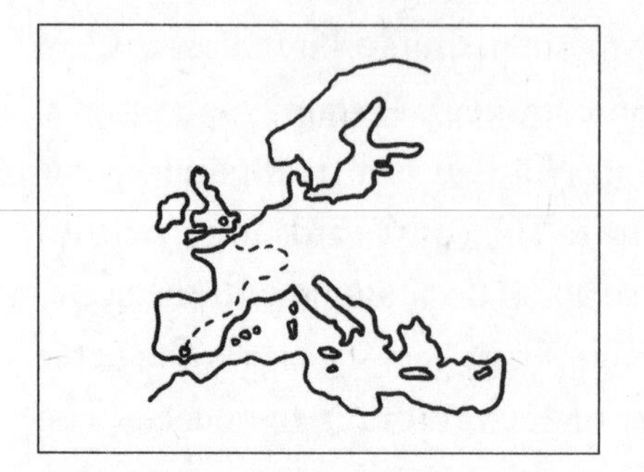

CHAPTER 15

EUROPE

AS OUR JOURNEY AROUND the world of games approaches its terminus, London, we arrive in Europe for a visit to the place that many regard as the modern-day Mecca of games: Germany.

69: Spiel des Jahres

HOLLYWOOD HAS THE OSCARS. Musicians have the Grammys. Scientists have the Nobels. Mathematicians have the Fields Medals. But if you are a game developer, then the ultimate accolade is the Spiel des Jahres, or "Game of the Year."

The prize has been awarded since 1979 when it was first given to a game called Hare and Tortoise, based on the Aesop's fable. To qualify for the prize, games must have been released in the German-speaking world. While that first winner was originally an English innovation, in recent decades Germany has proven itself the hub of modern-day board game development led by games like the Settlers of Catan, which won the award in 1995. But the German gaming revolution was already well under way before the prize brought the industry into the spotlight.

In a country trying to distance itself from World War II, the aversion to war games with names like PanzerBlitz acted as a catalyst for a completely new strand of gaming. German law prohibited the import of war-related toys. Conquering was out. Constructing was in.

Germany has a toymaking tradition going back to the toy makers of Nuremberg in the fourteenth century. The town has long been celebrated as the toy capital of the world. Today the international toy fair in Nuremberg is the highlight of the toy developer's year. Although the main focus was on wooden toys, by the end of the nineteenth century board games began to make an appearance. In 1884 Bavarian publisher Otto Robert Maier decided to augment his book and pamphlet production with the publication of a new board game: A Trip Around the World. Based on Jules Verne's book, the printed board came with beautiful tin figures to race around the globe. (I hope that Maier would appreciate my own journey around the world in eighty games.)

So successful was Maier's version that he ditched the books and dedicated himself to making more games. The company that he founded is now synonymous with the great tradition of board games coming from Germany: Ravensburger. It is the Pixar of board games. By 1902 he already had over one hundred games in his catalogue. The economic hardship in Germany and the exodus of Jewish toy makers following the rise of the Nazis led to the collapse of the manufacture of toys. But the spirit of gaming in Nuremberg couldn't be suppressed, and in 1950 the town once more opened its arms to the toy makers of the world.

Because Ravensburger didn't own the rights to the big-name board games like Monopoly, Scrabble, and Cluedo, it had to concentrate on developing games of its own. A team of in-house game developers was created to help the company compete with Waddingtons and Parker Brothers. The creation of the Spiel des Jahres was one of the industry's tactics to stimulate the German market, and it worked.

The award of the prize can really lift a game from obscurity into the limelight. Once adorned with the rosette on the box marking them as prizewinners, games fortunate enough to be so honored can end up selling as many as thirty million copies worldwide. Tom Werneck, one of the regular judges and the founder of the prize, explained their goal when they created the award: "Our vision was to put such a strong pressure on the game industry that they struggle to bring the highest possible level of quality on a competitive market. When we look back now, 30 years later, we see that it worked out because the award *Spiel des Jahres* had such a strong influence on sales figures that it worked like a sting in their flesh inflaming competition."

The criteria that judges are looking for in awarding the coveted prize are originality and creativity in stimulating a new sort of game play; also the game shouldn't be impossible to learn, so the rules need to be clear and comprehensible. They reward the layout of the game, the design, and the aesthetics of the game as whole. The judges also seem to be suckers for a good story.

It is notable that game creators are not called inventors in Germany but authors. The German public looks out for the latest Reiner Knizia or Wolfgang Cramer game like readers seek out the new John Grisham or Stephen King book. It isn't good enough just to have good game mechanics. It has to be dressed up in a good storyline. Take the Quacks of Quedlinburg, which won the Connoisseur-Gamer Game of the Year in 2018. (Frankly I'd give it a prize just for the name!) Each year the town of Quedlinburg holds a festival in which each day apothecaries and quacks cook up potions to compete against each other to prove they are the greatest potion mixer in the land. The game is all about cooking up potions without their exploding in your face. This has become another of my family's favorites.

If you look at the front of the box, you will find the name of the creator of the game: Wolfgang Warsch. It is rare to find the name of the game developer so prominently featured on games coming out of the English-speaking world. The importance in Germany of celebrating the game's "author" can be judged by the publication in 1988 at the Nuremberg Toy Fair of the "Coaster Proclamation." This was a statement signed by leading game developers that they would not publish games with any company that refused to put their names on the box fronts. The proclamation was originally signed on the back of a beer coaster as the developers sat round drinking. But it worked. Since then the practice of celebrating the "author" on the front of the box has become standard.

70: Pandemic

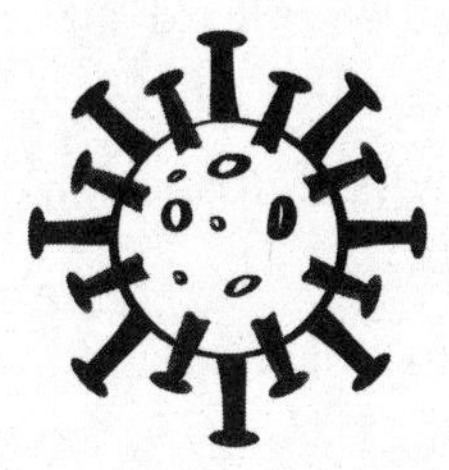

PERHAPS GIVEN WHAT THE PLANET has been through in the last few years, some readers will be less than keen to read a section titled "Pandemic." But bear with me. Because the game of Pandemic, released in 2008, represents a very interesting moment in the evolution of games. As a game where you compete not against each other but collectively against the game itself, it is a great example of the sort of innovation that the Spiel des Jahres is trying to stimulate.

The designer of Pandemic, Matt Leacock, created the game after playing board games with his wife. "Early in our marriage, we played all sorts of games together. Some worked well for us, while others... didn't. I recall a particularly disastrous negotiation game where the emotions generated by my psychological manipulation, relentless competition and betrayals during play bled out into the real world and put a strain on our relationship."

He began to wonder if it was necessary for all games to be competitive. Perhaps he could come up with a game where he and his wife worked together. Then they came across a game created by master board game creator Reiner Knizia. Knizia has created over seven hundred games and won multiple prizes for his games. Based on *The Lord of the Rings*, this was a game where the players work together as a group of hobbits to defeat Sauron and destroy the ring.

Leacock recalls the sensation of playing a game where you are all on the same side: "When playing our first cooperative game together, we had a markedly different experience. The challenges were all the more exciting for being shared—and regardless of whether we won or lost, we both enjoyed the ride because we were in it together."

Knizia's cooperative game offered an exciting new direction for board games. Instead of playing each other, you played the game! Leacock set out to add his own contribution to this innovative style of game and came up with Pandemic. The board depicts a map of the world, and at the beginning of the game, four virulent diseases have broken out across the globe. The players have to work together to come up with cures for the diseases and treat the outbreaks before they overwhelm the world. Now considered a classic in the genre of collaborative games, Pandemic has sold over five million copies worldwide.

I love Pandemic because it is a game for the nerds at heart. Rather than portraying characters wielding swords or guns in an attempt to take over the world, you get to play a research scientist wielding a syringe and a test tube. Each player takes on a different role, but each offers different skills in the attempt to overcome the pandemic sweeping the board.

Leacock is very aware of the important emotional dynamic that a game should elicit as you play it. "You really want people's emotions to go up and down, you don't want a flat line. There are certain things you can do to get people excited or afraid, and just modulate that emotion throughout the course of play. Also, building up to a certain crescendo and then having some sort of resolution, but with greater stakes as the game progresses up to that climax near the end, some sort of resolution at the end."

When you are playing a game against others, the competition between players is responsible for providing this emotional roller coaster. That's why a good game never allows one person to run away with winning the game early on and allows all players to stay in the game to the end.

But when you are playing together, you need to design a game where the emotional highs and lows emerge out of the game itself. You don't want a game to be too easy or so difficult that you give up. Pandemic manages this dynamic remarkably well. As the cards turn over and cause a sudden surge in the infections across the world, the players are challenged to respond to this new scenario, planning their strategy together and pooling resources to beat the game.

One challenge with creating cooperative games is making sure that one bossy player doesn't just tell all the others what to do. You really don't want the game to descend into a game of solitaire where the other players are just following the instructions of the lead player. This difficulty aside, the genre of cooperative games inspired by Knizia's and Leacock's work has continued to produce a slew of games further working out this type of game play.

As we have seen over the course of our journey, collaboration between different players in a game is not a new idea. Games like pachisi or bridge involve players working in pairs to beat the other team. War games like Diplomacy or Risk can see temporary alliances form, which ultimately are betrayed once a player makes a break for the winning line. Dungeons & Dragons requires a team to work together to create a satisfying adventure, with the Dungeon Master serving more as facilitator than adversary. The prisoner's dilemma proved that players do best if they can overcome their primal instinct to do what is good for them as individuals and work collaboratively. Still, games like Pandemic are unique in that all players are collaborating to beat the game.

One of the more recent innovations that Leacock is exploring with game developer Rob Daviau is a game that develops and grows each time you play it. Playing a game again shouldn't feel like *Groundhog Day*. They had the idea that instead the current game should depend on how you played the previous games. Called Pandemic Legacy, the game models our passion for watching boxed sets of a TV series over many nights rather than simply watching a film in one evening. They even call each new version of Pandemic Legacy a new season. The idea of an extended narrative that spans multiple game-play sessions is a striking innovation for the world of gaming.

But more pressingly, Leacock believes that games have the capacity to effect social change and hopes to create games that can engage people politically. As Leacock commented in the *New York Times* in the middle of the Covid-19 pandemic in reference to the role a board game might play in combating a global health crisis, “Board games have comparatively low stakes, but I’ve learned they have much to teach us: We all need to play to our strengths, balance short-term threats against long-term goals and make sacrifices for the common good. If we can communicate, coordinate and cooperate effectively we might better overcome this uncaring, relentless and frightening opponent.”

71: The Best Board Game Ever

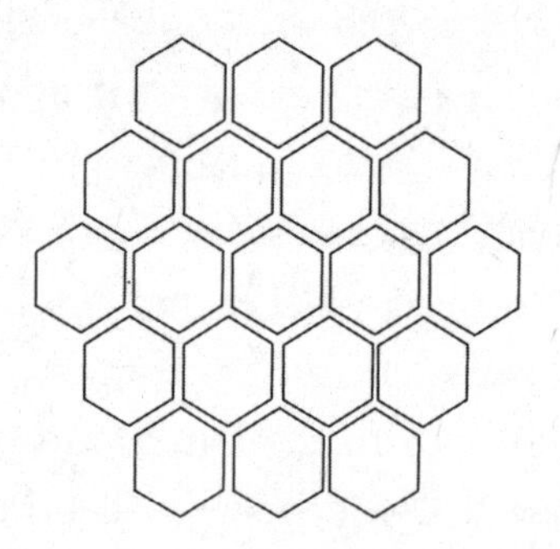

A TITLE LIKE THIS IS likely to elicit a huge range of responses and disagreement. It’s like saying a certain book is the best one ever. Impossible. But there is one game that does float to the top quite often when you ask people to choose.

Many years ago I was a guest on a phone-in radio show here in the United Kingdom. It was a three-hour late-night show where I was billed as the presenter’s friend, there to discuss a whole range of topics that were current for the day. I’d done it a few times before, and it was always interesting, touching on subjects as varied as the role of the monarchy in modern Britain and the use of drugs in sport.

One of the liveliest sections we had on the radio show was an hour dedicated to finding the best board game ever. Was it Risk? Or Ticket to Ride? Certainly not Monopoly. Among the public phoning in there was a clear winner. At the top of the list of board games they’d take with them to a desert island was the Settlers of Catan. First published in Germany in 1995, it has now sold twenty-two million copies in thirty languages. This game is credited with the modern-day resurgence of board games that has swept the world in the last few decades. Sales have continued to rise year after year, and many cities around the world are home to thriving cafés dedicated to playing board games, like my local venue Draughts in Hackney.

Board games prior to the Settlers of Catan often had a military or imperialist theme to them. It was about using your army to wipe out the opposition and take over the world. But in post–World War II Germany, such narratives did not resonate well. Indeed, it was not possible to publish a game with a picture of a tank on the front. This led developers to explore more constructive narratives on which to base their games.

The Settlers of Catan was created not by a career game designer but by dental technician Klaus Teuber. He was a big fan of Viking mythology, and so his idea for a game was a story about settlers discovering a new island and then foraging for goods in order to build a new settlement there. The element of constructively building a new land broke the cycle of war games. The other element that makes the game work so well is the trading that is allowed between players for different resources. You might have lots of wool but need ore to upgrade your settlement to a city.

Anyone prepared to trade wool for ore? That social component in the game dynamic allows the varied characters of competitors to play a role in the evolution of the game.

The Math of Catan

What got me most excited when I opened my first box of Catan was the board—or, I might say, boards, because the game essentially has a different board each time you play the game. The board is made by piecing together nineteen hexagons in a beehive-like lattice.

The number nineteen might seem curious, but this is the third number in a sequence called the hex numbers, which starts 1, 7, 19, 37, 61... We saw these numbers already when we were laying out counters on the carrom board in India. This is also the number of pieces you need as you build out an island made of hexagons with a single hexagon at the center, surrounded by six more hexagons and then another twelve to make nineteen hexagons in total. As we saw in the section on the carrom board, a mathematical formula will tell you the number of hexagons you need if you want to play a massive expansion of the game. With n concentric hexagons you'll need

$$3n^2 - 3n + 1$$

smaller hexagons.

The smaller hexagons in Catan have different qualities as they generate different resources from which to build the settlements. There are four fields, which yield grain; four forests, which yield wood; four pastures on which to rear sheep for wool; three mountains from which to mine ore; three hills from which to extract brick; and finally one barren desert tile. With these nineteen hexagons, how many different boards can you make? Are you likely to get a new board each time you randomly place the tiles?

Think of an empty grid of nineteen hexagons. We are going to fill the grid from the center. There are nineteen choices of which tile to put in the center. Now move to the next empty hexagon (we number them in a spiral out from the center). There are eighteen choices now for which tile to put in this position. So that's already 19 × 18 possibilities. As we move around the spiral choosing different tiles from the remaining pile, we end up with 19 × 18 × 17 × ... × 2 × 1—written as 19!—possibilities.

But there is a complication because these tiles aren't all different. If I take a board that I've laid out and swap the four field tiles around, the board doesn't really change. There are 4 × 3 × 2 × 1 = 4! ways to swap these four field tiles around. I need to factor in the rearrangements of all the tiles that represent the same resource. This cuts down the number of possible boards. We now have

$$\frac{19!}{4!4!4!3!3!}$$

boards. We divide by a 4! for each of the four field, four forest, and four pasture tiles and by a 3! for each of the three mountain and three hill tiles. That's 244,432,188,000 different boards. If the dinosaurs had started playing the Settlers of Catan as soon as they'd evolved, then at a game per night they would not have got through all the boards possible.

Actually, a few further twists that we haven't factored in increase the number of boards even further. First, number counters get laid down in the tiles to determine the probability that the tile will produce a resource when two dice are thrown. The game has a set configuration for these number counters, but the arrangement can be laid on top of our board in six different ways by rotating the configuration by a sixth of a turn around the central hexagon. Second, the nineteen hexagonal pieces sit inside a coastline on which there are nine ports into which you can assign nine trading ship pieces (five different, four the same). So there are 9!/4! ways to arrange these. So now we have a total of

$$\frac{19!9!\times 6}{4!4!4!3!3!\times 4!}$$

different boards. To add a third and final twist, a mirror image of a board arrangement isn't really different; nor is a rotation of the board. This cuts the total down by a sixth.

So, how can mathematics help you play and win the Settlers of Catan? A useful little trick may help out. At the beginning of each turn, you throw two dice to determine which tiles generate resources. Each tile has one of the number counters we laid down with numbers from two to twelve on it. If the number gets thrown by the dice, then the tile generates a resource. Every player with a settlement on that tile gets the resource. This is an interesting new feature of the game; usually only those whose turn it is benefit from a dice roll, but now anyone might benefit from your throw. This means that players stay engaged even when it is not their turn.

But recall from our analysis of Monopoly, with two dice there is only one way to get a two or a twelve, but there are five ways to get a six or an eight. In fact the number counters even have dots on them to help you understand the change in probability. The twelve counter has one dot; the eight counter has five dots. So some tiles will be generating resources more frequently. Since you receive a resource if you have a settlement on a corner of that tile, here is a trick to work out where is it best to place a settlement. A settlement sits on the corner of three tiles. Add up the dots in the counters on each of these three tiles. You want this to be as high as possible so that you are increasing the chances of getting resources.

One more feature of this game has intrigued me as a mathematician. You build roads along the edges of the hexagons. You score points if you have the longest road. But what is the longest possible road that you could make? You are not allowed to go over a road twice. This is one of the questions where Leonard Euler's Bridges of Königsberg, which we encountered playing dominoes, is going to be useful. In the land of Catan, every junction has two or three vertices coming out of it, so according to our analysis in Königsberg, you're only going to be able to cover two of them, except for the point where you begin and end. There are fifty-four vertices in total. You are going to be able to cover two of the edges out of these vertices.

They are counted twice. That gives fifty-four edges plus the extra edge you get for the start and end point, which counts for one extra edge. So a road length of fifty-five could be possible.

Such a road will never occur in a game because other players' settlements will get in the way. But it's the sort of question I can't avoid contemplating when I see such a setup. Probably one of the best strategies for playing games against me is to include some irrelevant mathematical angle that will distract me while you concentrate on winning the game.

Although modern European games are moving away from warfare, one game that came out of ancient Europe was notable for being perhaps one of the first games simulating war. And it came from a civilization that took its warfare very seriously.

72: Ludus Latrunculorum

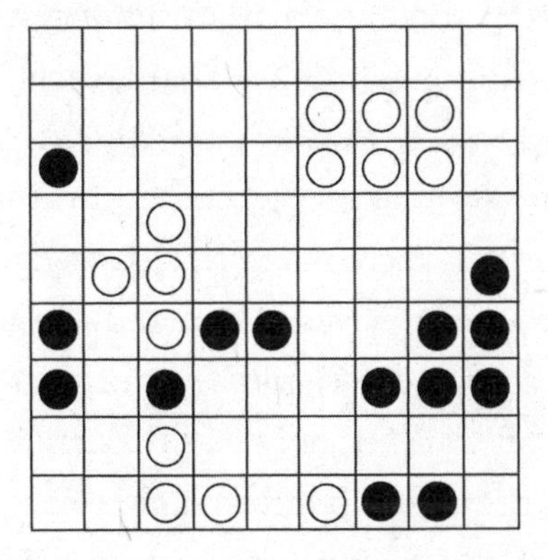

BECAUSE OF MY NAME, Marcus, I have always felt something of an affinity for Roman culture. In Latin lessons at school, we would always use my name to decline second declension nouns, and any dialogue in our textbooks would invariably feature a Marcus, whose lines I was required to read. I used to love visiting Roman remains across Britain and then, in later life, throughout the Middle East, imagining a previous Marcus who might have guarded the Roman encampment. So it was a great thrill to discover that like me, the Romans loved their games.

Roman soldiers would carry gaming boards in their packs despite the extra weight just so they could carry on gaming as they marched through Europe and the Middle East. They loved throwing dice and playing games of chance. But the Romans were also responsible for creating probably one of the first games that simulated strategy in warfare, a game perhaps to train soldiers in military tactics.

Called Ludus Latrunculorum, it is known as the "game of little soldiers" or the "game of mercenaries." It is another game where the rule book has been lost to time, but the plethora of boards that exist across Europe testify to the popularity of the game. The board consists of a square grid of varying dimensions. There are examples of 7 × 7, 7 × 8, 8 × 8, 8 × 9, 9 × 9, and 9 × 10 grids. Grids were often carved into stone slabs or roof tiles, but others have been found etched into the steps of the Parthenon in Athens and the Basilica Julia in Rome.

Hemispherical counters made of glass, bone, or pottery found buried next to boards support the idea that players each had an army of stones that would be moved across the board to simulate two armies engaging in battle. On the most popular, 8 × 8 board, each player seems to have had at least sixteen pieces, as in chess, although they might have played with as many as twenty-four pieces each. Although we have no rule book, quite a few written accounts of the game give us some inkling as to how it might have been played. Ovid and Martial both have verses and epigrams dedicated to descriptions of the game in action.

One source reveals that not only was this game played in the comfort of the home or camp, but public matches were staged with crowds gathering to witness the masters of the game. Among these masters was celebrated player Roman senator Gaius Calpurnius Piso, who lived in the first century AD. A poem titled "Laus Pisonis" celebrating the senator describes his prowess at the game: "But what player has not retreated before you? . . . In a thousand ways your army fights: one piece, as it retreats, itself captures its pursuer: a reserve piece, standing on the alert, comes from its distant retreat—this one dares to join the fray and cheats the enemy coming for his spoil. Breaking the opponent's defensive line, it may burst out on his forces and, when the rampart is down, devastate the enclosed city . . . and both your hands rattle with the crowd of pieces you have taken." This final image depicts Piso's hands shaking all the pieces he has taken in the face of his opponent. Unfortunately his skills in the game didn't translate to success in life.

A conspiracy that he led to overthrow the emperor Nero was thwarted, and Piso was forced to take his own life as punishment for the betrayal.

From the poem describing the game play, it is clear that this was a dynamic game, with pieces moving across the board rather than staying static like the stones in the game of Go. This was not passive territorial occupation but active warfare. Apparently there wasn't a set starting position for the pieces; rather each player could lay out their own formation as the opening part of the game. A tomb uncovered in Chichester in 1996 did, however, contain a game laid out as if a few moves had been made; it hints that the blue and white pieces were initially lined up at the back of the board as in chess. It does appear to share with games like Go the idea that stones are captured when they are surrounded. As Ovid writes in *The Art of Love*, "One piece falls before a double foe." But other lines in the poem hint that pieces have the chance to escape if the player can protect their piece or attack the attackers in a countermove. There are shades of John Conway's Game of Life at play in the way counters live and die.

The Roman game of war is one of pure strategy, though another European war game from the twentieth century combines a very winning blend of strategy and chance, creating one of the most popular modern-day war games.

73: Risk

INVENTED IN 1957 BY French filmmaker Albert Lamorisse, Risk is a game of world domination, played out on a map of the world divided up into six continents and forty-two countries. Players populate the countries with armies and then challenge neighboring countries occupied by other players to a battle fought with dice to expand territory. I love Risk not just because it meets my criteria for a game, with a perfect blend of strategy and luck, but because it is full of fascinating mathematics.

One decision at the outset, as you lay out your armies, has to do with which continents you want to concentrate your forces on. During the game, on each turn you generate new army pieces that you can deploy depending on the number of countries you occupy. But you get bonus armies if you occupy a whole continent. Here we see mathematics at work to create a balanced points system. A big continent like Asia, comprising twelve countries, will reward you with more bonus armies if you succeed in occupying every country, relative to a smaller continent like Australasia, which only has four countries.

To decide which continent to go for, you want to simply calculate the number of bonus armies you get divided by the size of the continent to determine the continent that gives you more bonus armies for your buck. Here are the results.

Continent	Number of Countries	Bonus Armies per Turn	Bonus Armies per Country
Asia	12	7	0.58
North America	9	5	0.55
Europe	7	5	0.71
Africa	6	3	0.5
South America	4	2	0.5
Australia	4	2	0.5

Europe immediately sticks out as the continent to go for. But be careful. The topology of the board balances out the perceived advantage of taking Europe. Just as with the iconic map of the London Underground, the physical geography of the world map of Risk is not important compared to how the countries are connected together. In the London Underground map, I don't care what the distance is between Caledonian Road and Holloway; I just want to know that there is a connection between the two. In Risk a continent is more vulnerable to attack if there are more connections to other continents.

Here is the real map of Risk that you need to analyze. Like the London Underground map, it ignores the sizes of the countries and the distance between them and simply represents the fact that two countries share a border and therefore have the potential to attack each other.

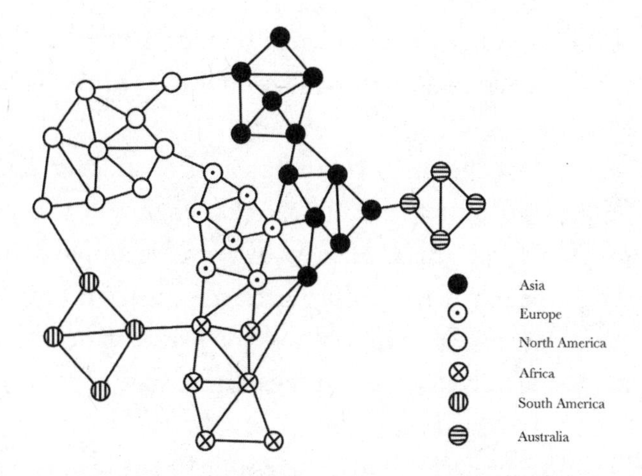

Europe has four countries vulnerable to attack and eight different invasion routes, while Australasia has just one vulnerable country with a single invasion route. That makes Europe much harder to defend. The continent in the sweet spot here is North America, generating a good proportion of bonus armies while still being easy to defend, with only three vulnerable countries and three invasion routes.

Risk illustrates beautifully how the mathematics of point scoring and the topology of networks are key to creating a carefully balanced gaming experience. Understanding them is key to planning an optimal strategy as you set your armies on the board. Less settled is what you should do next, since mathematical opinion is at odds over whether attack or defense is the best strategy as the game plays out.

There is a nice bit of asymmetry at work in the way the dice work to determine a battle between two countries. Each army in the country gets a die, but there is an upper limit. The attacking country gets to use up to three dice compared to a maximum of two dice for the defending country. When they are thrown, the dice are placed in order and paired off to determine the outcome of this round of the battle. The attacking army needs to beat the throw of the defending army to knock it off the board. Otherwise the attacker loses its army. For example, suppose the attacking country has three armies and throws five, four, and one, and the defending country has two armies and throws six and three. Since the defender's six beats the attacker's five, the attacker loses an army for the first pair of dice, but three is beaten by four, so the second pair of dice sees the defender lose an army. Play continues until one country's armies are wiped out.

So who has the edge in this dice game? The attacker? Or the defender? Should you sit and soak up attacks, or should you go on the offensive on each turn? This is another beautiful example of how Markov chains can help a player to navigate the expected outcome. The essential condition for applying this bit of mathematics is that each round of the battle only depends on the current state of the armies, not on how they got to this state.

The key construct in the mathematics of Markov chains is the Markov matrix, a rectangular array of numbers encoding the probabilities of different outcomes. The Markov matrix for Risk requires calculating the probability, given the number of armies in each country, that each country will lose armies. The rows and columns of the matrix correspond to the number of armies owned by each player. The entries of the matrix record the chances of going from one state to another. The first mathematical analysis of these probabilities, made in 1997 by Turkish mathematician Baris Tan, resulted in the Markov chains predicting a slight advantage in the long run for the defending armies. However Tan made a classic error in probability theory: assuming that two events are independent when they aren't.

If two events are independent, then the chances that both happen is calculated by multiplying together the probabilities for each independent event. The chance of getting a six on one die is $\frac{1}{6}$. If you throw two dice then one die doesn't affect the chances of the other die landing on six. The throws are independent. So the chances of getting two sixes is $\frac{1}{6}\times\frac{1}{6}=\frac{1}{36}$. But in Risk things are more complicated because you are comparing dice across two different throws. The probability that my top die beats your top die is dependent on whether my second die beat your second die. If my second die was a six, then there is a $\frac{5}{6}$ chance that my top die will beat your top die, because my top die must also be a six. But if my second die is a one, then the chance my top die beats your top die is different.

This subtle dependence is very important when assessing probabilities not just in games but in real life. When Sally Clark was brought to trial in 1999 for the death of her two children, her defense was that they had both died of cot death. The prosecuting lawyers argued that there is a 1 in 8,500 chance of one cot death occurring; they then multiplied these probabilities together to get the chances of two cot deaths occurring as 1 in 8,500 × 8,500 = 72,250,000. The numbers seemed overwhelming, and Clark was convicted. But the defense lawyers argued at retrial that the math was wrong. If one child dies of cot death, then that actually might massively increase the chances of a second cot death occurring in the family, either due to genetic or environmental influences. Clark was finally released in 2003.

When, in 2003, Jason Osbornbe put the new probabilities into the Markov matrix for Risk, the balance of power shifted. Taking into account this subtle dependence across throws, the new probabilities reveal that there is a slight advantage for the attacker in battles of Risk. The mathematics now supports an aggressive approach to playing the game.

With all of this aggression and global conquest, it's not hard to recognize that many of the authors of the games we have been discussing are male. Though the gaming industry has traditionally been very male dominated, that is beginning to change. In fact, women game designers have proven that they are not averse to creating a good game of war.

74: L'Attaque and Women in the Gaming Industry

I STUMBLED ON THE FRENCH GAME of L'Attaque at All Souls College in Oxford. In a corner of the common room is a box of games for fellows to play of an evening after the port has been passed. The pile of games sits just under a portrait of Lawrence of Arabia. I can imagine L'Attaque being a game that he would have enjoyed playing, and I wonder, given the vintage of the game, whether it might even have been donated by him when he was a fellow.

The game plays like a mix of chess and Battleship, and its aim is to capture the opponent's flag. All pieces are hidden from view, as in Battleship, and include scouts, mines, sappers, and a spy, made from cardboard but beautifully illustrated to indicate each soldier's rank. The game is played on a 9 × 10 board with two regions of water that cannot be occupied by any piece. When you move a piece in front of a square occupied by your opponent, you shout, "Attack!" and the piece with the highest numerical value survives. If you attack your opponent's mine, then you get blown up unless you are a sapper, in which case you successfully defuse the mine. You each have a spy whose sole aim is to capture the commander in chief. All other pieces will capture the spy. It's endearingly described on the box as "war without tears."

But the surprising discovery I made is that the author of the game is a woman—a rarity during this period of game development. The game was originally designed by Mademoiselle Hermane Edan in the 1880s, but she only received a patent for the game in 1909. It was translated and distributed in the English-speaking world in the 1920s but still seems to have maintained its French name. The game pieces were printed at a premises in Aldersgate Street in London until the Germans hit the building with a bomb in 1940 and ended the game for a while.

Anyone who has played Stratego might recognize the game play. Stratego emerged in the Netherlands during World War II. A Canadian pilot who had been shot down in 1942 was being hidden from the occupying German army, and to alleviate the boredom, the pilot showed his protectors a game that he made with pieces representing two armies facing off on a board. The game was so much fun that it inspired the Dutch hosts to register the game under the name Stratego.

And yet Stratego seems to have a lot in common with L'Attaque.

That is not to say that L'Attaque was the first game to depict war in board game form in this way. Edan's original game shares quite a lot in common with an even older Chinese game called Dou Sho Qi, or "the jungle game." Here, instead of armies, the pieces represent animals of differing rank. The elephant is the most powerful, while the rat is at the bottom of the pecking order. Unlike in L'Attaque, the location of the animals is known by each player. It is a game of complete information. The boards bear significant similarities, however, with two areas of water marked on the board in similar locations. The rat, although at the bottom of the pile, is able to swim across the water. A game even closer to L'Attaque emerged in China; it is called Luzhanqi, or "land battle chess," and the ranks of pieces are hidden. Considering that Edan's uncle was first consul for France in Shanghai, it is possible that he could have brought one of these Chinese versions home to France, inspiring her own creation. We can never know for sure.

Following the success of L'Attaque, Edan continued to create new games. She submitted a number of them to the famous Concours Lépine, a competition in France designed to encourage innovation that is still active today. The games included L'Attaque and two more: Naval Combat and The Faithful Bird.

The original game is a rare example during this period of a game made by a woman, rattling the prevailing perception that gaming is an inherently male activity. Increasingly many players in board game cafés and online are women, but the industry still suffers from few women following Edan's lead and breaking into the making of games.

Bringing more women into the gaming industry increases diversity as well as the number of interesting new games emerging. Take ten people from the same background, brought up on the same games, and you'll probably get a similar idea from all ten. Alternately, consider American designer Elizabeth Hargrave, an avid bird watcher, who won the Connoisseur's Spiel des Jahres in 2019 for her wonderful game Wingspan. It is a card-based game about bringing birds into a nature reserve. Hargrave's passion for nature is the inspiration for many of the games she has made. "There were too many games about castles and space, and not enough games about things I'm interested in. So I decided to make a game about something I cared about."

She got into board games on a skiing trip. As she had grown up in Florida, skiing wasn't her specialty. Fortunately a friend brought along some board games, which caught her imagination while she was passing the time on the trip. She really enjoyed the mathematical side of the games, exploring different strategies for how to win Settlers of Catan or Ticket to Ride. It was that puzzle aspect of gaming that she enjoyed when she started to make her own games, especially the challenge of producing game mechanics that would overcome certain faults that emerged in early versions of games she was developing. But at some point she decided to place a moratorium on playing any more games with castles. As she explained in an interview in *Slate* magazine,

> Publishers are targeting "typical gamers" with their themes and mechanics and their big, heavy games about whatever fantasy creatures that I don't care about. And then it becomes this feedback loop where they're making games for this very niche market of people, and then the people who become board game designers are coming out of that market, and then they're making more games just like that. You miss out on the diversity of people and the diversity of games that different kinds of people would make.

Another up-and-coming female game designer is Gemma Newton, owner and founder of Moonstone Games based in the United Kingdom, which produced the successful farming game Plotalot. Nature again is the inspiration for Newton's game, which taps into her joy of growing vegetables in her allotment, hence the name Plotalot. The dynamic of trying to successfully harvest vegetables while keeping the slugs at bay gave her the feeling of playing a game in the garden. Newton's card-based game captures that competition with nature.

In the game you're up against other farmers competing to grow vegetables. Trouble is there isn't enough land for all of you, and you're battling the constant bane of gardeners: bugs of different shapes and sizes. Despite its rather idyllic setup and beautiful artwork, the game can get quite vicious, tapping into a well-know game mechanic called "take that," where your moves directly attack an opponent's progress toward victory. Like Wingspan, Plotalot brings fresh ideas by looking to nature rather than obsessing over goblins and castles. In Newton's words, "The balance of male to female players is extremely good, as is the balance to young and old. However, what I don't see as much is female designers or publishers. This is an area where I feel women can add real value in terms of creativity and fun. Women have a different take on gaming."

I have already introduced you to Brenda Romero, who powerfully used a game to explain the slave trade to her daughter. Another very striking game that she developed in 2009 shows the power new perspectives can bring. In Train, players compete to move passengers represented by yellow figures in trains to reach their destination in the most efficient manner. It is only toward the end of the game that you discover the destination is Auschwitz. The game is meant to get players to question how easily people are prepared to follow rules. Once again, Romero has used games as a powerful vehicle for discussing emotionally difficult issues like the Holocaust or the slave trade.

The gender differences when it comes to playing games are rich and complex and can't simply be assigned to obvious stereotypes. Boys like weapons. Girls like flowers. A study of Taiwanese gamers revealed that women played to feel a sense of achievement and were looking for social interaction. For the men, gaming was a way of passing the time. In France research showed that women enjoyed competition and the challenges of a game, while the men used games to release stress. Although today the gender split between players is pretty equal, the companies creating games are still dominated by men. But that is beginning to change.

While Hermane Edan was designing the war game of L'Attaque, a contemporary in France was creating one of the great pencil-and-paper games of all time.

75: Pipopipette or Dots and Boxes

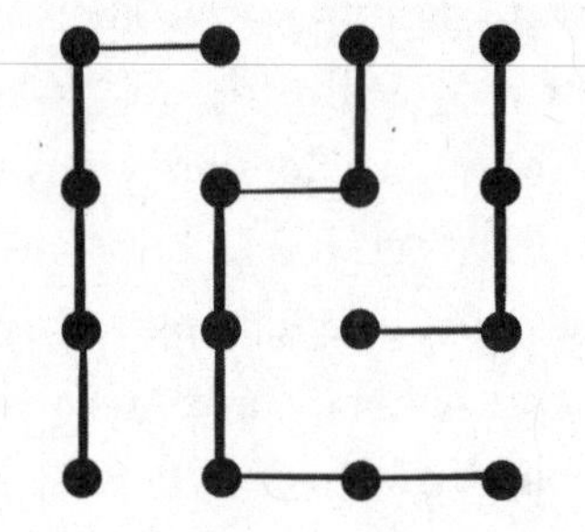

TRAVELING ROUND THE WORLD with kids in tow, being stuck on trains for twenty-four hours from Ürümqi to Kashgar or on buses from Mumbai to Kochin, I've become familiar with a good range of games that just need pencil and paper to play: Hangman, Battleships, Sprouts. One of my favorites is Dots and Boxes—or Pipopipette, as its inventor, Eduard Lucas, named it when he published the game in 1895.

I was particularly struck by his name. Could he be the same Lucas whose work on the theory of numbers I'd studied as a research mathematician? Indeed he was. Lucas came up with a clever way to test whether special sorts of numbers of the form $2^n - 1$ studied by the seventeenth-century French mathematician Marin Mersenne are prime or not. Lucas used his method to prove that $2^{127} - 1$ is indeed prime. It remains the largest prime number ever discovered without the aid of a computer. The biggest primes discovered today with a computer are still exploiting Lucas's clever trick for testing primality.

As well as prime numbers, Lucas also loved games. The Tower of Hanoi, more a puzzle than a game, was invented by him. The puzzle consists of three wooden rods with a stack of discs of different sizes on one rod. The challenge is to move all the discs one at a time to another rod, with the rule that a larger disc cannot be put on top of a smaller disc. How many moves does it take? It turns out that with n discs, the number of moves is the Mersenne number $2^n - 1$ that Lucas had studied.

When the toy was marketed, a colorful story was added to the game. Legend had it that an Indian temple in Kashi Vishwanath housed a huge version of the puzzle with sixty-four golden discs, and the Brahmin priests were tasked with moving the discs from one post to the other. It was claimed that once the task was complete, the world would end. If the legend were true, and the priests were able to move discs at a rate of one per second, using the smallest number of moves, it would take them $2^{64} - 1$ seconds, or roughly 585 billion years, to finish, much longer than the predicted lifespan of the sun. It seems like the legend was made up by Lucas to promote the game, although Kashi Vishwanath was the site of a legendary challenge set by Shiva to test Vishnu and Brahma. The challenge had nothing to do with moving golden rings, although Shiva does have sixty-four different forms, which might have been Lucas's inspiration for the choice of sixty-four rings.

Lucas also introduced the wonderful game of Pipopipette, or Dots and Boxes as it is known in English. Dots and Boxes starts with a rectangular grid of dots. Players then take turns drawing lines between pairs of horizontally or vertically adjacent dots, with the aim of forming boxes. A player captures a box by completing its fourth line and initialing it, then must draw another line. After all lines on the grid have been filled in, the player who has captured the most boxes wins.

What is important in this game is the number of long chains where you can claim three or more boxes in one turn. If you are left with a board with an even number of long chains, then there is a strategy for the first player to win. If there is an odd number, then the second player controls the board. So the strategy for each player is to create odd or even numbers of chains. If the board is a 4×4 grid of dots (resulting in a 3×3 collection of boxes), then the second player can always force an odd number of long chains and hence can always win the game.

To play this game well, it is worth learning a strategy for forcing your opponent to give you a long chain. It's called the double-cross strategy.

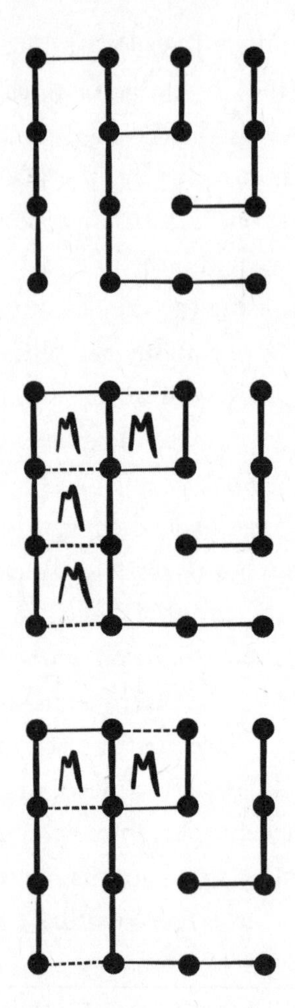

Suppose you are faced with situation 1. There are two chains: one of length four, the other of length five. Your opponent has just given you the length-four chain in the hope that you will claim it and then be forced to give them the length-five chain. That is what the scenario in the middle diagram describes. But there is a smarter move: the double-cross strategy. You take all but two boxes and draw a line to leave a length-two chain. Now your opponent has to end up giving you the long chain of length five. They can claim the length-two chain but then are forced to put in the line that lets you claim the length-five chain.

As the board gets bigger, it becomes increasingly more difficult to determine who has the winning edge. The largest board that has been determined to date has 4 × 5 boxes in it. This gives an even number of boxes, and optimal play will always result in a draw. But for bigger boards it is much harder to analyze who might win, which makes it a perfect game for a long train, bus, or boat trip on a journey around the world.

Lucas dedicated the game to his students at the École polytechnique in Paris, who loved playing it. Alas, Lucas died tragically in a strange culinary accident. During a banquet celebrating the advancement of science, a waiter dropped a plate, and a piece of the smashed crockery slashed Lucas's cheek. These wounds became infected, and he died some days later of sepsis at age forty-nine.

The next game is another one created by a French mathematician, and its global popularity is further proof of mathematicians' great skill as game developers.

76: Dobble

MATHEMATICS PLAYS AN IMPORTANT ROLE in playing and winning games, but it is also a powerful ingredient when it comes to inventing them too. Many games are based around stories. A bunch of explorers settling a new land. An army conquering the world. A group of adventurers searching for treasure. Farmers growing vegetables. But some games shun the story and just go for the joy of the game mechanics to excite players. One of my favorite examples of this is a game that is basically a beautiful piece of mathematics in disguise.

The game is called Dobble in the United Kingdom or Spot It in the United States. It consists of a circular tin box containing a set of fifty-five cards. Each card has eight symbols. Any two cards from the pack will always have one, and only one, symbol in common.

The basic version of the game gives each player a card, and the rest of the cards sit in a pile with one card showing on top. The first player to identify the symbol that is common to their card and the card on the top of the pack claims the card and places it on top of their card. A new card is revealed and the players compete again to claim this card by spotting the shape that is common to their card. The winner is the one who collects the most cards once the pack has been cleared.

The origins of the game go back to a French enthusiast of mathematics called Jacques Cottereau. In 1976 he produced a set of thirty-one cards with six insects on each card in a way that each pair of cards shared an insect. He'd always been fascinated by a classic mathematical puzzle called Kirkman's School Girl Problem. In 1850 the Reverend Thomas Penyngton Kirkman submitted a puzzle to the *Ladies and Gentleman's Diary*. "Fifteen young ladies in a school walk out three abreast for seven days in succession: it is required to arrange them daily, so that no two shall walk twice abreast." This was a sort of human sudoku puzzle.

The mathematical geometry that solves Kirkman's School Girl Problem is also the key to creating Cottereau's game of insects. Cottereau simply enjoyed playing the game with his family, but when his daughter's brother-in-law Denis Blanchot saw the game, he recognized its potential. He expanded the underlying geometry to create a game with fifty-five cards, each containing eight symbols such that each pair of cards shared a symbol. In 2009, Dobble was released in France.

Given that there are only eight symbols on each card, at first it feels like a bit of magic that any two cards share one and only one symbol. Of course there are some silly ways that you can arrange this. For example, I could put the symbol of a cat on each card and perhaps add some other random symbols to confuse the situation, but once you've realized that this is how the cards are made the game becomes rather boring.

A slightly more extreme version might be to have a symbol for each pair of cards. But given that each card would be paired with fifty-four other cards, you'd need fifty-four symbols on each card.

There are $\frac{55 \times 54}{2}$ = 1,485 pairs in total, so that would mean coming up with 1,485 different symbols. Possible but not terribly efficient. Dobble gets away with eight symbols on each card chosen from a bank of only fifty-seven symbols. So how does it manage this? Not magic but mathematics.

The key to being able to create such an efficient pack of cards lies in geometry. Classical geometry consists of lines and points. Two points lie on a unique line. If the line corresponds to a symbol and the points correspond to the cards, then two points lying on a unique line already translates into the idea of two cards having a unique symbol on each.

The geometry Dobble exploits is not the usual Euclidean geometry that we are used to drawing on the graph paper in our exercise book but an example of something called a *finite geometry*. To explore this, we will look to a mini version of Dobble with seven cards. Here is a picture of the geometry. It is called the *Fano plane*.

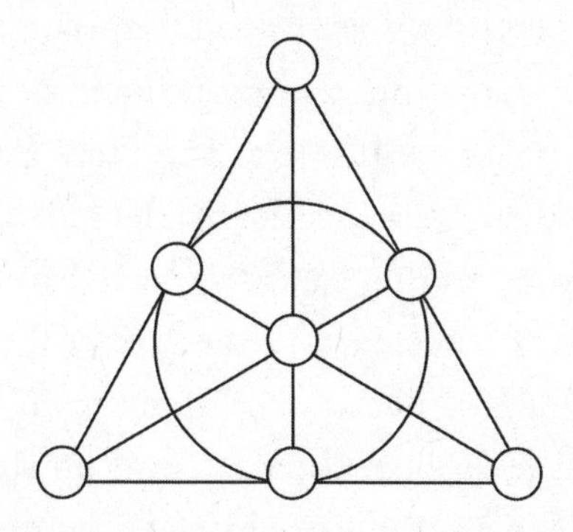

The geometry consists of just seven points. There are seven lines that connect the points. One of these lines looks like a circle, but this is still a line in this geometry. In our mini version of Dobble, I place a card on each of the seven points and associate a symbol with each of the seven lines, and then if a card is on a line, then it gets that symbol.

Each card has three lines through it, so there are three symbols per card.

Remember that in any geometry, two points have a unique line through them. This means that whatever two cards we pick, there is a unique line through those cards that translates into a unique symbol matching on both cards.

If you prefer to build your geometries through coordinates and numbers, there is another way to describe the Fano plane. It's a kind of digital geometry where the coordinates are zeros and ones. There are three coordinates that can be filled with zeros and ones, but we exclude the point (0, 0, 0). This gives us seven different points. Take any two points, then the third point on the line through these two is got by adding the coordinates together and interpreting $1 + 1 = 0$. For example, the line through (1, 1, 0) and (1, 0, 1) goes through $(1, 1, 0) + (1, 0, 1) = (1 + 1, 1 + 0, 0 + 1) = (0, 1, 1)$.

Having made a mini version, let me show you the geometry behind Dobble. While the Fano plane is based on zeros and ones, essentially the geometry of Dobble is based on the prime number seven. Arrange forty-nine blank cards in a 7×7 array. Let's number the cards (a, b) for a card in the ath row and the bth column. I am going to draw lines through these cards. Each line will correspond to a symbol that gets drawn on the card it passes through. The lines are going to come in batches of seven parallel lines. The first batch of lines is easy to describe. It's just the lines corresponding to the seven columns.

The second batch of lines starts at the card at the top of the column but now passes through the next card along from the card below. So if the card starts at $(1, b)$, then the next card on the line is $(2, b + 1)$. As I move down the grid, the line shifts a column. If you get to the end of a row, then you just come in at the beginning of the row on the next step down. So here is a line through the third card in the top row.

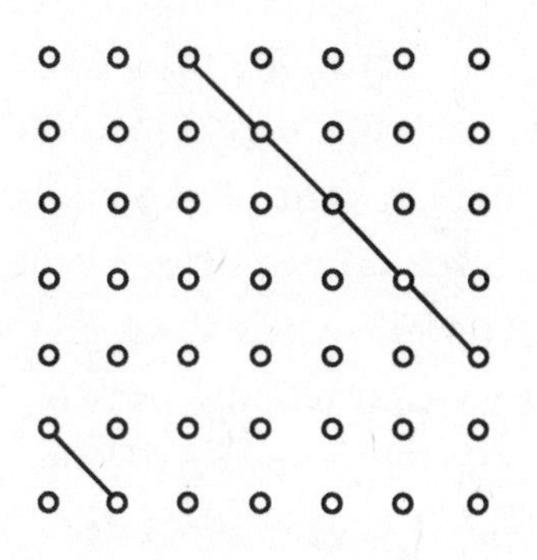

This gives seven parallel lines corresponding to each of the points at the top of the columns.

The third batch just shifts two cards along as it descends. So the line from $(1, b)$ goes to $(2, b + 2)$ and carries on like this, again reentering the grid on the left if it exits the grid on the right.

I can keep on doing this until the seventh batch consists of lines shifting six along so $(1, b)$ goes to $(2, b + 6)$. Finally there is an eighth batch of seven lines, which is the lines corresponding to the rows. That's a total of $8 \times 7 = 56$ lines. Each line has its own symbol, which gets painted on the cards it passes through.

Since there are eight lines through every point, each card has eight symbols on it. But hold on. We only have forty-nine cards, and the Dobble pack has fifty-five. Where are the other points coming from?

Take the batch of parallel lines that we have created. Currently they don't meet. But what if I create a point at infinity where they meet and place a card there? It's a bit like parallel lines in a Renaissance painting meeting at what we call the *vanishing point*. So eight batches of parallel lines give rise to eight points at infinity, which each have a new card placed there. And I can introduce a new line (like the horizon in the painting), which passes through each of these new cards. I've added another line (and symbol) to give fifty-seven lines and eight new cards to make $49 + 9 = 57$ cards in the game.

But hold on. Aren't there only fifty-five cards in a Dobble box? Indeed, there are. Strangely enough, the makers of the game have not included two extra cards that could be added to the pack. Why did they drop two cards? The explanation isn't clear, but one suggestion is that a conventional pack of cards comes with fifty-two cards and two jokers plus a marketing card, making fifty-five in total. This is because the cards are often printed on a 5 × 11 grid. One thought is that the company making these new cards just simply repurposed the equipment for printing a conventional pack of cards to print Dobble sets. So the machine only had space for fifty-five out of the fifty-seven possible cards. There are fifty-seven symbols in a Dobble set so all the lines are there, but the two missing cards mean that the geometry of Dobble is missing two points. It doesn't affect the game play, but it does upset the mathematical aesthetic of the game, at least in my mathematical eyes.

Dobble Geometry

The geometry at the heart of Dobble and the Fano geometry are examples of geometries called *finite projective planes*. These are geometries that satisfy the following axioms:

- Two points are connected by a line.
- Two lines intersect in a point.
- There are four points, so no line contains more than two of them.

We say that a projective plane has order n if each line has $n + 1$ points and each point is on $n + 1$ lines. The Fano plane is a projective plane of order two. The geometry behind Dobble is of order seven. Mathematicians know that for every prime number p, there are projective planes of order p. We can also cook up projective planes of the order of a power of p, like $3^4 = 81$. But if the order is not a prime power, then we don't know. In 1991 it was finally proved that it is impossible to make a finite projective plane of order ten. It is still open whether you can make a finite projective plane of order twelve.

Again, if you prefer your geometries in coordinate form rather than pictures, then in Dobble the coordinates are got by choosing numbers from zero to six. We again have three coordinates, but the first coordinate is either one or zero. If it is zero, then we have the further convention that the next coordinate is one or zero. If it is zero, then the last coordinate must be one. This gives 49 + 7 + 1 = 57 points.

To draw lines through points, we use something called *modular arithmetic* or *clock arithmetic*. This is essentially the idea that you do calculations on a clock with seven hours marked on it from zero to six. So if you want to add four and five, the answer isn't nine but two, which is where you end up on a seven-hour clock if you start at four and add five hours.

This is actually what is happening when the lines exit the right of the grid and reenter at the left. For example, the line from (1, 5), which steps two along, is basically adding the coordinate (1, 2) each time. So the line that starts at (1, 5) then goes to (1, 5) + (1, 2) = (2, 7) = (2, 0). (In this arithmetic seven is the same as zero.) It then goes to (2, 0) + (1, 2) = (3, 2), then to (4, 4), (5, 6), (6, 1), (0, 3), and then back to (1, 5).

There is something rather thrilling about playing Dobble with my kids. While they may think they are picking out the crazy clown common to both cards, I know that they are actually identifying the unique line that passes through points in the projective plane of order seven. There is another very successful game that pushes these finite geometries into hyperspace.

77: SET

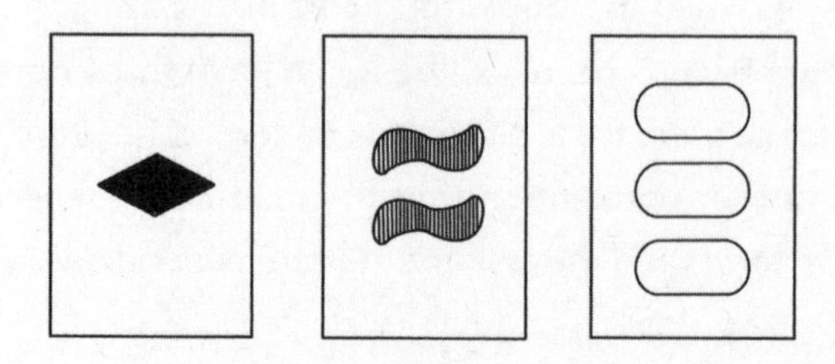

THE GAME SET IS BASED on a geometry in four dimensions, although the geometry was not the initial inspiration for the game. It was invented in 1974 by population geneticist Marsha Jean Falco when she was studying epilepsy in German shepherds. She decided to represent the genetic data on the dogs by drawing symbols on cards and then searching for patterns in the data. The pattern searching gave her the idea for the game. She used to play the game with her family, who enjoyed it so much that finally her children persuaded her to share the game with the world. SET was finally released to the public in 1991, going on to win over thirty-five awards for best game.

In this game you are looking to pick three cards to form a set. Each card has symbols on it, but this time the symbols that appear have one of three different shapes: ovals, squiggles, or diamonds. They come in one of three different colors: red, green, or purple. The shapes are shaded in one of three options: empty, striped, or solid. Finally, the number of shapes that appear varies from one to three. So there are four attributes (shape, color, shading, and number), each with three different possibilities. So that means the box of SET contains $3 \times 3 \times 3 \times 3 = 81$ different cards.

Three cards form a set if, when you consider the attributes in turn, they are either all the same or all different. For example the three cards shown at the beginning of this section form a set because the shapes, shading, and number are all different, while the color (black in my greyscale version of the game) is the same.

Here is another set:

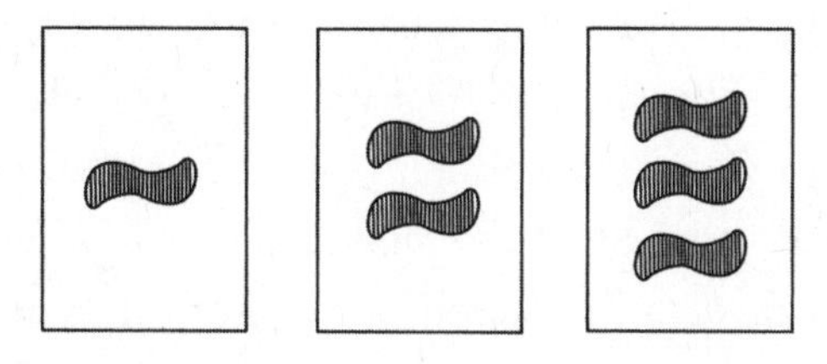

Same shapes, same colors, same shading, different number.

This however is not a set:

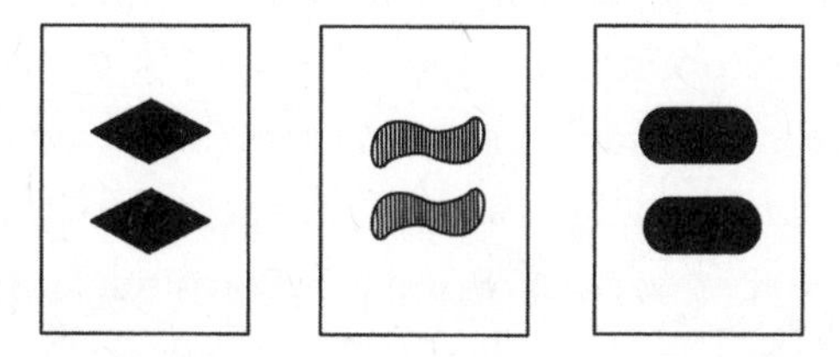

The shading is neither all the same nor all different.

The game proceeds by laying out twelve cards so that players stare at them, trying to find three that make a set. For example, can you find a set in these twelve cards? There is more than one. It's actually quite tricky to spot a set. But once a player finds a set, they claim those three cards, and three more are laid down. The winner is the person with the most cards at the end. Boys out there should be warned though: research suggests that women are quicker at spotting sets than men.

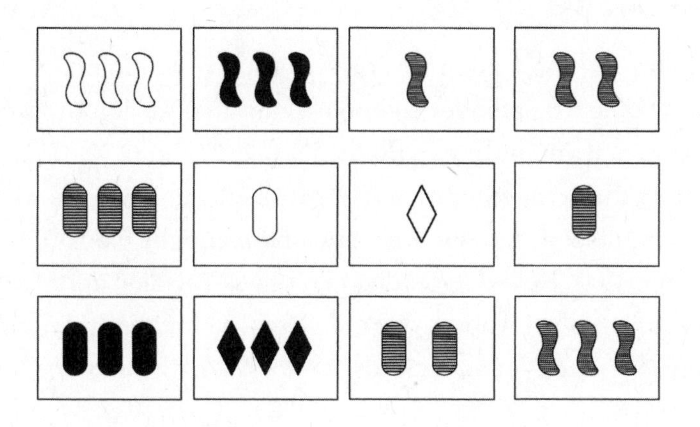

A lot of questions that arise in SET can be answered by understanding that, once again, this game is actually about the mathematics of a finite geometry in disguise. This time the geometry is four-dimensional, corresponding to the four attributes, and each coordinate is a number from zero to two. It is called the *four-dimensional affine geometry over the field of three elements*, or AG(4, 3) for short. The points in this geometry correspond to the cards, and a set consists of three points that lie on a line. Notice that if you have two cards, you can always find a unique card that completes the two cards to make a set. This translates into the fact that a line through two points in the geometry always intersects in a third point in the geometry.

Since we are stuck in a three-dimensional universe, it is hard to draw pictures of these higher-dimensional geometries. But if we create a game of SET that ignores color and just keeps track of the other three attributes, then the game consists of $3 \times 3 \times 3 = 27$ cards that make up a three-dimensional geometry. These cards can be thought of as the twenty-seven tiny cubes that make up a Rubik's Cube, although a Rubik's Cube doesn't actually have the hidden twenty-seventh cube at the center. Now we can see how two cards always have a third card that makes up a line in this geometry. Choose any two cubes in the Rubik's Cube, and there is always a third cube that lies on the line joining the first two cubes.

Making a Set

To find the third point on a line through two points, you add the two coordinates for the points and then divide by two. For example, if you have two points (2, 1, 2, 2) and (2, 0, 0, 2), then adding these together gives (1, 1, 2, 1). This time we are working with modular arithmetic with three hours on the clock 0, 1, 2. So $2 + 2 = 4 = 1$. How do you divide by two? Dividing zero and two are easy, but what about one? Well, find a number that when you double it gives you one. Since $2 \times 2 = 4 = 1$, dividing one by two

gives the answer two. So the third point on the line is (2, 2, 1, 2). Notice that this works to give a set because looking at the three points—(2, 1, 2, 2) and (2, 0, 0, 2) and (2, 2, 1 , 2)—either the coordinate is all the same or all different.

Unlike in Dobble, where there is always a symbol shared by two cards, it isn't guaranteed that every twelve cards will always have a set. The probability that a set does not appear in twelve cards turns out to be about 3.2 percent. But this estimate is only derived from computer simulations on the game. No one has proved a precise formula for the chance. If twelve cards don't contain a set, then three more cards are added to the grid. It's still possible that these fifteen cards won't contain a set. In fact there is a possible configuration of twenty cards with no set among them.

A set is only guaranteed when you have twenty-one cards on the table. This you can prove by turning SET into geometry and showing that twenty-one points must contain a line. The maximum number of sets you can have in twelve cards is fourteen. This corresponds to the fact that twelve points in this geometry can have a maximum of fourteen lines running through them.

Like in Dobble, although players of SET think they are simply spotting patterns in the cards, the truth is that we are all taking an adventure around a geometry in four-dimensional space. That's the power of a game.

78: Mornington Crescent and Nongames

AROUND THE WORLD IN EIGHTY GAMES

WHILE THE REFORM CLUB marked the beginning and ending of Phileas Fogg's journey around the world in eighty days, the British Museum probably represents a good place to tie the loop of my own journey. It was, after all, the place where I first saw the Royal Game of Ur that inspired my lifelong adventure collecting the games of the world. As I approach the end of my journey, returning to the center of London, I will probably use the London Underground at some point to make my way back to the British Museum, where my adventure started. As I enter this final leg of the journey, it seems an ideal moment to share a strangely British game based on one of the London Underground stops: Mornington Crescent. This stop on the Charing Cross branch of the Northern Line between Euston and Camden Town has given its name to what is known as a nongame.

The point about a nongame is that as the game is being played, it appears that there are rules governing the next move, and yet, as an outsider watches the game, it seems impossible to discern what the governing rules are. Mornington Crescent first appeared as part of a radio panel show on the BBC called *I'm Sorry I Haven't a Clue.* First broadcast in April 1972, the show is basically a parody of radio and TV game shows and has run for decades on BBC Radio 4. In the opening episode of the sixth series broadcast in August 1978, the new game of Mornington Crescent was introduced.

In the game panelists take turns naming stations on the London Underground. The winner of the game is the first person to name Mornington Crescent. As the game was played, it seemed like some rule determined the permitted stations that could be said next. Was it that the next station had to be on the same line? Or that you could get to it in a set number of stops? When a panelist finally announced Mornington Crescent, the other panelists would congratulate them on a game well played. The whole thing was a mystery since it was unclear what the rules were. The point was that there were none.

The game evolved out of tension that had emerged in the production team. The panelists had fallen out with one of the senior producers on the show, who had become very unpopular. Sitting round the pub after one of the recordings of the show, one of the panelists spotted the producer heading to their table to join them. "Quick, let's invent a game with rules he'll never understand." Mornington Crescent was born.

The game turned out not to be new. It was a version of a game that had been invented by mathematicians in 1969. Theirs had been based on another station: Finchley Central. The description of the game appeared in an article in the mathematical magazine *Manifold* about the idea of nongames: "Two players alternate naming the stations of the London Underground. The first to say Finchley Central wins. It is clear that the 'best' time to say 'Finchley Central' is exactly before your opponent does. Failing that, it is good that he should be considering it. You could, of course, say 'Finchley Central' on your second turn. In that case, your opponent puffs on his cigarette and says, 'Well, . . . shame on you.'"

It's interesting that the panelists of the game show chose the rather curious station of Mornington Crescent as the destination for their game. It is not a station many find themselves visiting because of its slightly obscure location. Most people get off at Camden Town, the neighboring station, which is only a couple of minutes away by foot. It has strange opening times restricted to rush hour often. It was in fact closed for many years for refurbishment. The unpredictable access to the station fueled the feeling of strange rules governing its role in the network. There is also something funnier about the name when compared to Finchley Central, which the comedians recognized.

What is so intriguing about this game is that it isn't trivial how to play it. As the authors of the original game explained, you could just blurt out Finchley Central on your first move. But that defeats the object, which is to find a way to continue the game. That is in some sense the winning move. But ultimately a good game is one that satisfyingly seems to resolve with someone saying Finchley Central or Mornington Crescent at just the perfect moment.

My own favorite nongame is Numberwang, a mathematical game show that appeared on the UK comedy sketch show *That Mitchell and*

Webb Look. The contestants, Simon and Julie, call out numbers until the host, played by Robert Webb, suddenly declares, "That's Numberwang." As with Mornington Crescent, the rules are completely unexplained and appear to make no sense to the onlooker. Simon can announce a number deemed wrong by the host only for Julie to offer the same number and be greeted with the winning "That's Numberwang." One reason the sketch is so successful is that it taps into many people's experiences of math lessons at school. The recurring sketch got more and more elaborate as the show went on. I particularly liked the German version Nümberwang and the advert produced for the board game version you can play at home, which included a board, numbers, two four-hundred-sided dice, and a thirty-seven-volume rule book. Memories of wrestling with Dungeons & Dragons came to mind.

A good gaming friend told me about another rather curious nongame called The Game that he started playing with friends some years ago. The aim is to forget about The Game. If one day, while sitting on the tube, you suddenly remember The Game, it means you've lost the latest round. You then have to contact all your friends to tell them, "I've lost The Game," at which point everyone remembers that they are playing. The Game is then reset, and everyone goes back to trying to forget about it. "Speaking of which, I've just lost The Game again, so I'll have to let them know." Sorry, Theo!

The nongame of Mornington Crescent works partly because it captures the point of all games: that you create artificial obstacles that mean the game doesn't end before it's started. A game of snakes and ladders could just involve moving your piece from the first square to the last square, and yet we set up rules that mean you have to find your way to the end at the behest of the dice. Good games have endings, or moments of resolution of some kind, much like listening to a piece of music. The point isn't to hear the last chord. It's to be taken on a journey that leaves you feeling transformed or content or enlightened by the time you reach the end. But generally you do want an ending, a moment when you can applaud the journey you've all been on collectively as listener and performer.

If you ever fancy playing Mornington Crescent, here are some useful game theory tips on how to play and win. First off, why is blurting out Mornington Crescent at the first opportunity a bad move? There are several critiques of this as a winning move. If you always do this as your move in the game, no one will play with you. Also it reveals that declaring Mornington Crescent earns a greater payoff the more moves it takes before you say it.

It's a bit like the game at the heart of the Saint Petersburg paradox. A coin is tossed. If it lands tails, you win $2, but each time it lands heads, the stake is doubled. What you want is a long streak of heads before the first tails appears so that your win is as big as possible. The paradox at the heart of this game has to do with how much you would pay to play this game. It turns out that your expected win playing this game is an infinite amount of dollars, which means you should play whatever the entry fee. And yet most people won't pay more than about $10 to play.

In Mornington Crescent there is a hidden payoff function at work that somehow measures how many laughs the panelist will get. It increases each time a station is mentioned. Let's suppose that, in keeping with the Saint Petersburg paradox, the number of laughs doubles each time. The game is now one of brinkmanship. What is the optimal moment to cash in? The same game is embedded in a number of ways auctions are set up. For example, in a Dutch auction, the price of the object on sale gradually decreases. The first person to make a bid then claims the object at the current price.

There are only 272 stations on the London Underground, and one of the rules that does seem to be adhered to in the game is that you aren't allowed to say a station's name twice. So if you are playing and you find that 270 stations have been named, then it obviously makes sense to choose Mornington Crescent on your next go. But your opponent realizes this must be your move, so they will declare Mornington Crescent on the 270th round. Since you have realized they are going to do this, it makes sense to declare Mornington Crescent on round 269. But once again your opponent realizes you have to do this and so will preempt this with a call of Mornington Crescent on round 268.

You probably see where this is going. Given a finite number of stations, mathematical logic implies that in a competitive game against another opponent, you must declare Mornington Crescent on your first round. The only trouble is that the payoff is as small as it could be.

The whole experience has a feel of the prisoner's dilemma where competition results in a bad payoff for both. What if players cooperate? Now one could go through all the stations and the player going second can end with Mornington Crescent on round 272, winning 2^{272} laughs, which are shared by both players. But there is still a feeling that the laughs aren't exactly shared. Surely the person going second benefits more. Won't player 1 perhaps deviate from the plan at some point to claim a big, if not the biggest, win? But won't player 2 be considering the same? And perhaps because Mornington Crescent must be declared on round 272, it doesn't quite have the same surprise as if it is declared unexpectedly earlier. Does the value of the laugh decrease at some point as we head toward the inevitability of hearing Mornington Crescent? The dynamics of this game are actually much subtler than one might have first expected. As someone once commented on the game, parodying Dr. Johnson's take on London, "She who is tired of Mornington Crescent is tired of life."

79: Infinite Games

> "There are at least two kinds of game. One could be called finite, the other infinite."

SO WROTE JAMES P. CARSE, and I was immediately intrigued. An infinite game? What is that? I must admit that playing Monopoly sometimes feels infinite, but I guessed that wasn't what the author of the 1986 book *Finite and Infinite Games* had in mind. Carse, a professor of philosophy and religion at New York University, wanted to explore two different approaches to the game of life. One focused on life enriched by past achievements and a sense of winning the game. The other preferred to focus on the possibility of continual renewal and future enrichment, of keeping the game of life going. On the one hand, life through the finite lens, on the other, life as an infinite game.

Given that my mathematics uses finite logical thought to try to navigate the infinite, I was fascinated to learn more about what constituted an infinite game and how it was played. The first difference is that you don't aim to win an infinite game. Winning terminates the game and therefore makes it finite. Instead the player of the infinite game is tasked with perpetuating the game—making sure it never finishes.

This means that rules must have the ability to change during the course of an infinite game. Players of an infinite game will move collectively to change the rules if they see that the current rules are likely to result in the game coming to an end and a winner being declared. This is a move in the infinite game: to change the rules to perpetuate the play. This is not acceptable in a finite game. "Finite players play within boundaries; infinite players play with boundaries."

The grammar of a language has some similarities to the rules of an infinite game. For a language to be able to continue and thrive and not die, the rules have to be able to evolve and change in order for it to keep track of the evolving world that it seeks to embody.

One of the interesting consequences of the freedom to change the rules is that deviancy is a quality of the infinite game player, something not tolerated in a finite game. Deviancy is often the key to culture evolving and not stagnating. These are the moments when new artistic movements emerge. Rule changes grow out of the rules of the past but offer something new to future players.

A finite game not only has a fixed end point but also has a beginning. An infinite game can have a beginning, but it could also have the quality that it is infinite at both ends, with no precise moment where the beginning of the game can be identified. Indeed, time is created within an infinite game rather than being measured by an external measure of time. It is its own world. It is impossible to say how long an infinite game has even been running.

An infinite game can have many finite games embedded in it, which players can win or lose, but these are just moves within the infinite game. Finite games have an audience that looks on as the game is played. They are theatrical in nature. Because the infinite game must keep the future uncertain, Carse believes that infinite play has the flavor of the dramatic rather than theatrical.

There is also a political angle to the tension between finite and infinite players. The infinite player is part of a collective social movement that works together to ensure the game continues. The finite player is an individual seeking the title of winner at the expense of the other players.

What binds both finite and infinite games together is the role that freedom plays. "Whoever *must* play, cannot *play*." To play games is to assert one's free will. You are voluntarily entering into the arena and submitting to the rules of the game. You cannot play a game if the rules are imposed on you. You freely choose to play by those rules. There is a democratizing role that games play in society. If not, it isn't a game you are playing. However, once you are within a finite game, you are submitting to those rules and in some ways relinquishing part of your freedom. In contrast, in the infinite game you maintain your freedom to act independently if necessary to ensure the game never finishes.

My journey around the world in eighty games is full of examples of finite games. So what is an example of an infinite game? Carse hints at examples that might satisfy some of the traits of the infinite game. Culture has the qualities of the infinite game. As soon as a cultural tradition seems to have become exhausted and run its course, players make a new move.

A Mozart or Rembrandt steps up to ensure that the game doesn't come to an end. The garden in contrast to the machine is another realm that illustrates the infinite game. The garden: self-sustaining, constantly evolving, never-ending. The machine: goal-oriented, consumer of external power, destined to decay.

Carse concludes his book with the rather cryptic statement, "There is but one infinite game." One realizes that we are all players in the infinite game that is being played out around us. If one were religious, which Carse was, I guess one might call it God's game. Or perhaps less controversially, nature's game. Certainly, it is the game of humanity. We may die as finite players, but our attempts to leave behind offspring, cultural changes, and scientific discoveries are part of our contribution to trying to perpetuate the game. But it is a game that we are very close to turning into a finite one, given the way we are treating our environment and our fellow players of the game. Let's hope that we can find a way for no one to win this delicate game we are all playing.

80: The Glass Bead Game

OUR JOURNEY IS AT ITS END, so what better way to conclude than with my favorite game. Invented by a German, it is a game no one knows how to play.

A few years ago I received an email from the BBC. "Hello. I'm the producer of a programme where we invite guests to choose 8 records, a luxury and a book that they would take with them if stranded on a desert island. The programme is called Desert Island Discs and we'd love to have you on the show." It was a very sweet email because it made no assumption that I knew what this program was. The show is one of the most iconic broadcasts in the United Kingdom, and I don't think I'm the only one who has been planning my list of eight records ever since I first heard the program at school.

The emphasis is mostly on the music guests choose. Mine was predominantly fantastic orchestral works with great trumpet parts like

Richard Wagner's *Parsifal* and Leoš Janáček's *Sinfonietta*. My luxury was my trumpet. I envisaged myself playing along to the discs as I sat on the deserted beach. But the book I chose to take with me on the island was all about a game. Set in the future, it's a fictional game whose rules are somewhat unclear. But when I read the book, I became obsessed with trying to play this game, and I think everything I've done since that moment has been an attempt to become a master of it.

The book, *The Glass Bead Game*, is the creation of German author Hermann Hesse. Players of the game are expected to synthesize themes from music, mathematics, history, linguistics, philosophy, and art, woven together almost like a story as the game proceeds. It is a game that plays with the total content of our culture. Hesse has a wonderful image of the Glass Bead Game player playing the game like an organist. "And this organ has attained an almost unimaginable perfection; its manuals and pedals range over the entire intellectual cosmos; its stops are almost beyond number. Theoretically this instrument is capable of reproducing in the Game the entire intellectual content of the universe."

The origins of the game are obscure but seem to go back to Pythagoras; they feature in ancient Chinese culture and then again in the Arabic-Moorish period around the growth of the house of wisdom. Gottfried Leibniz's dream of capturing philosophical thought in some universal language appears to be part of the development of the game. As time went on glass beads began to be used instead of letters, numerals, notes, or other graphic symbols to represent ideas.

The book is set at some point in the future where the playing of this game is undertaken by a monastic order whose members are chosen at a young age to be taken away from the trials and tribulations of real life so that they can train to play the game. The abstract nature of the state of the game meant that mathematicians were particularly skilled at playing it "with a virtuosity and formal strictness at once athletic and ascetic. It afforded them a pleasure which somewhat compensated for their renunciation of worldly pleasures and ambitions."

Throughout the book I've celebrated the way that mathematics is an integral part of making great games and finding cunning strategies for winning those games. But I think Hesse captures in his book that mathematics has a gamelike element in its own right. Because of the way mathematics often creates an abstract world outside time and place, it shares that sense of requiring the flight of imagination that is involved when you plunge into a game. Although mathematics does ultimately have utility, that is not the drive for those of us who dedicate our lives to its pursuit. We love doing mathematics for the sheer joy of playing the game. There is even a strange sense of winning in mathematics when you reach the final QED at the end of the proof, the checkmate, or grand slam of the mathematical game.

Each mathematician chooses the particular mathematical game that they feel matches their skills. It might be topology or geometry or differential equations. For me it was the game of group theory, a simple set of rules for understanding symmetry that, like the game of Go, could be played out so as to give rise to complex possibilities. From these simple rules emerge strange eight-dimensional crystals that dictate the fundamental particles of nature or the exotic 196,883-dimensional snowflake known as the Monster. The mark of a great game and great mathematics is the unending possibilities for new discoveries and surprises.

As I made my move to become a mathematician, to lose myself in math's abstract world, to join the monastic culture of the academic world of mathematicians, Hesse's description of those who had chosen to play the game felt like my story. But just as the game players did not restrict themselves to one subject, a narrowness that would have seen them failing in the game, I too was intent on pursuing my passion for music, for theater, for literature. It sounded like the game I wanted to play as I embarked on my life's journey.

The book follows the life of one of the great players of the game: Magister Ludi Joseph Knecht. The Magister Ludi is like the pope of the game, the highest, most revered position a player can attain. When I read the book as a student, Knecht was the character I aspired to. It's never made very clear throughout the book how the game is actually played, even if there is a winner. The game verges more on the playing of a ritual. It is a secular rational ceremony. And yet many games have their origins in something of capturing life in abstract ritual. From the moment I read the book as a student, I have been experimenting with what the rules of this particular game might be.

I recently reread the book, and I got something of a shock. My memory was of the joys of playing this game and how Knecht goes from being a student in school picked out by a senior player for his musical skills and then becomes one of the greatest players of all time. Either I had completely repressed the second half of the book, or it made no impression on me at the time. This documents Knecht's devastating recognition of the emptiness of playing the game for its own sake. Much to the dismay of his disciples in the game, Knecht forgoes his position and the sanctity of the world they inhabit and leaves to live a simple life in the world beyond the game.

I couldn't quite understand why I'd forgotten this twist. I guess at the young age at which I read the book, it didn't resonate. Today, having experienced similar moments of angst in my own mathematical life, Knecht's dilemma and choice felt very much the most important part of my reading this second time around.

Games are wonderful. But you cannot lose yourself forever in a game. It is important to leave the game and rejoin the real world around us. Maybe the game has helped one's journey through real life by providing a safe training ground to experiment with new ideas. Maybe it has simply allowed a period for the mind to rest and meditate on something other than the troubles of real life. Maybe it provides a new perspective. Maybe it is just a time out. But ultimately my rereading of *The Glass Bead Game* made me realize that putting the game back in the box and moving on is as important as that exciting moment when you first throw the dice or move a pawn or play a card at the beginning of the next game.

The Glass Bead Game therefore feels like an appropriate place to end our journey around the world in eighty games as we start to put away the pieces and return to life beyond the games.

CHAPTER 16

ENDGAME

PHILEAS FOGG managed to squeeze out an extra day over the course of his journey around the world. Every degree traveling east gained him four minutes, and the complete circuit of the globe of 360 degrees gained him a total of 4×360 minutes, or one full bonus day. And perhaps we, too, deserve an extra game to add to the eighty we've collected.

That eighty-first game is the book you are holding in your hand, capable of being read like a game, where the dice decide what you will be reading next. It is my effort to play Ludwig Wittgenstein's language game and understand what makes a game—the voluntarily erection of artificial barriers between us and our destination, which we then spend time trying to navigate.

Like a great book, a game allows us to step out of time and space and inhabit an alternative universe. But as this journey around the world has illustrated, our games are still very much linked to time and place. Our games are a window onto culture, a testament to history, a marker on the map. It is fascinating to witness how some games are reinvented multiple times across the world by different societies. There is something universal about throwing a die and racing round a track. And yet other games uniquely capture specific traits of a culture.

The aesthetic of games is really important too. There should be a beauty in a game—not just to the pieces and the board but to the mechanics of the rules and the narrative they give rise to, as simplicity gives rise to complexity.

Games are part of the art we have created as a species, allowing us to explore and share our inner worlds. They are a product of the development of consciousness in the human species. Because one is voluntarily submitting to the rules of a game, playing a game is an assertion of one's free will. We are not passive when we play games but actively involved, expressing our agency.

We have been playing games for thousands of years, and we have never grown tired of playing after all that time. Games are as popular now as they have ever been. People continue to gather in gaming cafés around the world, cracking open a box and setting up the pieces on the board. These are the modern-day campfires where people want to share in the experience of playing.

Business too is learning how to play. Everything is being turned into a game these days: education, fitness, sales targets. *Gamification* is the buzzword in the corporate world for taking advantage of people's love for competition and play. Change dry targets into winning points, clearing levels, completing missions, getting to the top of a leaderboard, and earning badges and suddenly you've engaged the emotions of your audience.

But setting business aside, the thing that has struck me time and again over the course of our journey is how much the playing of games overlaps with what I enjoy about mathematics. I love the challenge involved in solving a problem. That initial frustration of not knowing what is going on gets followed by the rush of adrenaline, the "aha" moment, as I see a way to the final QED. A game has that similar quality of trying to overcome barriers to get to the end. In both, the skills I have available and the difficulty of the challenge are the ingredients of a powerful cocktail that can reel me in for hours at a time.

We choose games that are challenging but not too challenging. We quite often are trying to achieve a sense of flow when we are playing a game so that we lose our sense of time outside the game and just live in the game's internal time frame. We want to get that rush of adrenaline at striving to reach the goal we've artificially set ourselves. If the obstacles we put up are too onerous, the game becomes work. If they are too easy, the game becomes boring.

John Cawelti, in *Adventure, Mystery and Romance*, characterizes the quality of this tension in literature: "If we seek order and security the result is likely to be boredom and sameness. But rejecting order for the sake of change and novelty brings danger and uncertainty... [T]he history of culture can be interpreted as a dynamic tension between the quest for order and the flight from ennui." I believe this applies equally to games and to mathematics.

Games share another important quality with mathematics: as with mathematics, you know when you've won or lost a game. That certainty was very alluring for me as a mathematician. I knew when I'd solved a problem. There was a right answer. Playing a game sets up a similarly controlled environment where there is a certainty in knowing what the outcome is. It is why artificial intelligence often begins with playing games as a challenge because we have removed the ambiguity inherent in most other practices of everyday life.

Games offer a beautiful, clean, unambiguous take on life that rarely matches the mess of reality. I think mathematics possesses a similar quality. Science is scrappy and quite often ambiguous, making it impossible to know for sure if you've got the right answer. Math has a beautiful, self-contained clarity about it. But our perfect mathematical models can often fall short in describing the way the messy world works.

I have discovered on my journey collecting games how deeply mathematics is embedded in the way we make and play games. I have encountered numerous examples of designers who have taken an interesting mathematical story and turned it into game. Look at how a very popular card game like SET works because players are exploring the geometry of a four-dimensional shape. The careful balancing act of scoring points in games taps into the way numbers can grow. Board games like Ticket to Ride and the Settlers of Catan are geared to maximize game play using mathematics. The mathematics helps reward those players playing a long game and resisting cashing in early. The mathematics of nontransitory relations is key to ensuring in a game like Dungeons & Dragons that no character in the group dominates the others. Rock beats scissors beats paper beats rock. Even a game like snakes and ladders works best when the mathematical arrangement of ladders and snakes is balanced. The making, as much as the playing, of a game benefits from a mathematical perspective.

So essential is mathematics to the way games are made that if you can tap into a mathematical mind-set, then you stand to gain an edge when you are playing games. Understanding the roll of a die can give you an edge in Monopoly. Mathematics can find you the best place to make your bet at the roulette wheel. Brownian motion dictates the ebb and flow in a game of backgammon. Binary numbers give you a strategy for playing Nim. Spotting patterns can give you an advantage in a game of Rock Paper Scissors. The best tracks to lay in Ticket to Ride or lands to settle in Catan are found by following the numbers.

People have often spoken about the close connection between mathematics and music. My journey around the world in eighty games has made me realize that there is an equally close connection between math and games. Johann Sebastian Bach's contemporary Lorenz Christoph Mizler described music as the process of sounding mathematics, and the concept may very well be extended to viewing games as a way of playing mathematics. Perhaps that's why the same game is reinvented in multiple places across the globe. It is tapping into the universal nature of the mathematics at the heart of the game. It's just that each society dresses it up in its own cultural story.

"Games are operas made out of bridges," declared Frank Lantz, director of the New York University Game Center. I love this description because it captures why there is such a close link between mathematics and games. Mathematics is the key to building bridges, but as Mizler articulated, it is also hiding behind the music we compose. Both games and mathematics combine the creativity and imagination of the artist with the logic and practicality of the scientist.

I've been once around the globe. I played eighty games, and I could just as easily keep on going: there are so many more games to play. We will keep on inventing new games. But as in every game, we must eventually reach the last square, the final card, the endgame—the finale where a winner is declared, the pieces collected and put back in the box. And so it is time to make my last move. All the chips on black six. K scores five. A four to win. Three in a row. Two points more. The ace of spades trumps all. And that's checkmate.

APPENDIX

How to Play Around the World in Eighty Games

TO PLAY AROUND THE WORLD IN EIGHTY GAMES, throw a die and move your counter to the game indicated by the roll. This is the first game to read. Place a mark next to the game to indicate that this square of the board has been visited. Roll again and move on to the next game. Keep rolling, reading, and marking off the games visited. When you reach the end of the book, just keep going, reentering from the beginning, imagining the board is circular. This time don't count the games that you have already visited. Keep going around and around the board until you have landed on all the games. How many different versions of the book will this game generate?

Around the Book in $6! \times 6^{74}$ Games

1	Backgammon	
2	The Royal Game of Ur	
3	Senet	
4	Rolling Bones	
5	Symmetrical Dice	

6	The Doubling Cube	
7	*Homo Ludens*	
8	Animal Games	
9	Language Games	
10	The Grasshopper's Games	
11	Chess	
12	Carrom	
13	Ludo	
14	Snakes and Ladders	
15	Ganjifa Cards	
16	The Buddha's Banned Games	
17	Hopscotch	
18	Chocolate Chili Roulette	
19	Nim	
20	The Ultimatum Game	
21	The Prisoner's Dilemma	
22	Go	
23	Chinese Chess	
24	Pick-Up Sticks	
25	Dominoes	
26	Mah-jong	
27	Zi Pai, Khanhoo, and the Origins of Playing Cards	
28	Whist	
29	Bridge	
30	Spades, Hearts, Diamonds, and Clubs	
31	Lady Charlotte and the Game of Parliament	
32	Tarot	
33	Hanafuda	
34	Pokémon Cards	
35	Dungeons & Dragons	
36	MangaHigh.com	
37	Cranium	
38	Mu Torere	

39	Cluedo	
40	Azad and *The Player of Games*	
41	Games and Riddles	
42	Theater Games	
43	Mozart's Dice Game	
44	Mexican Bingo	
45	Jogo do Bicho	
46	Adugo and Komikan	
47	Sapo	
48	Truco	
49	Perudo or Dudo or Liar's Dice	
50	Pitz, the Mayan Ball Game	
51	The Casino	
52	The Mansion of Happiness	
53	Monopoly	
54	Scrabble	
55	Wordle	
56	Rock Paper Scissors	
57	Ticket to Ride	
58	Prince of Persia	
59	Spacewar!	
60	Tetris	
61	The Game of Life	
62	Tic-Tac-Toe or Noughts and Crosses	
63	Mancala	
64	Gulugufe and Fanorona	
65	Achi	
66	Bolotoudou	
67	Nine Men's Morris	
68	Agram	
69	Spiel des Jahres	
70	Pandemic	
71	The Best Board Game Ever	
72	Ludus Latrunculorum	

73	Risk	
74	L'Attaque and Women in the Gaming Industry	
75	Pipopipette or Dots and Boxes	
76	Dobble	
77	SET	
78	Mornington Crescent and Nongames	
79	Infinite Games	
80	The Glass Bead Game	

If you use a die to choose a path around my book, then there are $6! \times 6^{74}$ different books you can generate, where $6! = 6 \times 5 \times 4 \times 3 \times 2 \times 1$. The best way to prove this is by induction. This is a clever way to prove that a formula works for all numbers. First prove it works for the smallest number possible. Next show that if the formula works for every number up to some point, then it works for the next number along. It's like the way I climb ladders even though I am scared of heights. I can climb onto the first rung. That's easy. If I've got to a certain rung on the ladder then taking one step up isn't too frightening. But now combine these two and it means I can climb as high up the ladder as I want. The same strategy works for proving a formula works for every number.

We shall prove that if there are $6 + n$ games, then the formula $6! \times 6^n$ counts the number of books you can generate with a six-sided die. If $n = 0$, then certainly a die can generate any permutation of the six books. So 6! is the correct formula. Now suppose that you have proved the formula for $n = k$. If I add a new chapter, so $n = k + 1$, then how many different books can I make? My die chooses six possible starting chapters. But then from here I am left with $n = k$ chapters to navigate, and my formula tells me there are $6! \times 6^k$ books that I can make out of these remaining k chapters. So in total that means there are $6 \times (6! \times 6^k) = 6! \times 6^{k+1}$ books.

So the number of ways to go around the world in eighty games is $6! \times 6^{74}$. That is a number with sixty-one digits. If you read a book every second since the Big Bang, then the number of books you could have read only has eighteen digits. So the chances are that the book you generate will never have been read before and will never be read again.

ACKNOWLEDGMENTS

I've been lucky to have been dealt a wonderful hand of cards to support me in the game of writing this book.

My Ace and King of Spades are my editors, Louise Haines at Fourth Estate and TJ (Thomas Kelleher) at Basic Books. They have expertly guided me up the ladders and spared me sliding down snakes as I put together the book.

The Queen and Jack of Spades are my copyeditor Jen Kelland and my line editor Brandon Proia who ensured that all my words and the order I put them in would be acceptable in a game of Scrabble.

The King of Diamonds is my wonderful agent Antony Topping at Greene and Heaton who does so much more than ensure I pass Go as many times as possible. Thanks also to my US agent, the Queen of Diamonds Zoë Pagnamenta, for looking after my interests in Atlantic City and beyond.

My Ace, King and Jack of Clubs: Theo Albert, Dudi Peles, and Rob Laidlow. Like any good game, it is important to beta test your ideas on players. Three great gamers whose conversations and feedback on manuscripts were invaluable as I battled the dragons and navigated the dungeons that make up the world of games.

Finally, I am lucky to have the Ace, King, Queen, and Jack of Hearts: Shani, Tomer, Magaly, and Ina. My Happy Family.

A fantastic hand of cards from which to attempt a grand slam.

BIBLIOGRAPHY

Websites

Board Game Geek: https://boardgamegeek.com
"Games" on wikiHow: https://www.wikihow.com/Category:Games
"Watch It Played" YouTube Channel: https://www.youtube.com/c/WatchItPlayed

Books and Papers

Allis, Victor (1988) A Knowledge-Based Approach to Connect-Four. The Game Is Solved: White Wins. Master's thesis, Vrije Universiteit.

Althoen, S. C. & King, L. & Schilling, K. (1993) How Long Is a Game of Snakes and Ladders? The Mathematical Gazette, Mar., 1993, Vol. 77, No. 478, pp. 71–76.

Antin, David (1968) Caxton's the Game and Playe of the Chesse. Journal of the History of Ideas, Vol. 29, No. 2, pp. 269–278.

Ascher, Marcia (1987) Mu Torere: An Analysis of a Maori Game. Mathematics Magazine, Vol. 60, No. 2, pp. 90–100.

Ash, Robert B. & Bishop, Richard L. (1972) Monopoly as a Markov Process. Mathematics Magazine, Vol. 45, No. 1, pp. 26–29.

Banks, Iain (1988) The Player of Games. London: Macmillan.

Barker, Joseph K. & Korf, Richard E. (2012) Solving Dots-and-Boxes. Proceedings of the Twenty-Sixth AAAI Conference on Artificial Intelligence (AAAI '12). Toronto: AAAI Press, pp. 414–419.

Bell, Robert Charles (1973) Discovering Old Board Games. Oxford: Shire Classics.

Bell, Robert Charles (1979) Board and Table Games of Many Civilizations. Rev. ed. London: Dover Books.

Benjamin, Arthur & Kisenwether, Joseph & Weiss, Ben (2017) The Bingo Paradox. Math Horizons, Vol. 25, pp. 18–21.

Berlekamp, E. R. & Conway, J. H. & Guy, R. K. (2001) Winning Ways for Your Mathematical Plays, Vols. I & II. 2nd ed. Natick, MA: A. K. Peters.

Binmore, Ken (2007) Playing for Real: A Text on Game Theory. Oxford: Oxford University Press.

Brown, G. G. & Rutemiller, H. C. (1973) Some Probability Problems Concerning the Game of Bingo. Mathematics Teacher, Vol. 66, No. 5, pp. 403–406.

BIBLIOGRAPHY

Buchin, K. & Hagedoorn, M. & Kostitsyna, I. & Mulken, M. V. (2021) Dots & Boxes Is PSPACE—Complete. In F. Bonchi & S. J. Puglisi (eds.) 46th International Symposium on Mathematical Foundations of Computer Science, MFCS 2021, Leibniz International Proceedings in Informatics, Vol. 202, pp. 25:1–25:18.

Burgiel, H. (1997) How to Lose at Tetris. Mathematical Gazette, Vol. 81, No. 491, pp. 194–200.

Caillois, Roger (1958) Man, Play & Games. Champaign: University of Illinois Press.

Carse, James P. (1986) Finite and Infinite Games. New York: Free Press.

Christie, Agatha (1936) Cards on the Table. Glasgow: Collins Crime Club.

Churchill, Alex & Biderman, Stella & Herrick, Austin (2019) Magic: The Gathering Is Turing Complete. arXiv preprint arXiv:1904.09828.

Consalvo, Mia (2005) Rule Sets, Cheating, and Magic Circles: Studying Games and Ethics. International Review of Information Ethics, Vol. 4, pp. 7–12.

Consalvo, Mia (2007) Cheating: Gaining Advantage in Videogames. Cambridge, MA: MIT Press.

Crombie, J. W. (1886) History of the Game of Hop-Scotch. Journal of the Anthropological Institute of Great Britain and Ireland, Vol. 15, pp. 403–408.

Dartigues, Jean & Foubert-Samier, Alexandra & Goff, Mélanie & Viltard, Melanie & Amieva, Hélène & Orgogozo, Jean-Marc & Barberger-Gateau, Pascale & Helmer, Catherine (2013) Playing Board Games, Cognitive Decline and Dementia: A French Population-Based Cohort Study. BMJ Open. Vol. 3. e002998.

Davis, B. L. & Maclagan, D. (2003) The Card Game Set. Mathematical Intelligencer. Vol. 25, pp. 33–40.

de Voogt, Alex (2015) The Role of the Dice in Board Games History. Board Game Studies Journal, Vol. 9, pp. 1–7.

Dell'Amore, Christine (2010) Prehistoric Dice Boards Found—Oldest Games in Americas? National Geographic News, December 15.

Demaine, E. D. & Hohenberger, S. & Liben-Nowell, D. (2003) Tetris Is Hard, Even to Approximate. In T. Warnow & B. Zhu (eds.) Computing and Combinatorics. COCOON 2003. Lecture Notes in Computer Science, Vol. 2697. Berlin: Springer.

Donkers, Jeroen & Uiterwijk, Jos & de Voogt, Alex (2003) Mancala Games: Topics in Mathematics and Artificial Intelligence. Journal of Board Game Studies, Vol. 6, pp. 135–148.

Dummett, Michael & Mann, Sylvia (1980) The Game of Tarot: From Ferrara to Salt Lake City. London: Gerald Duckworth & Co. Ltd.

Eerkens, J. W. de Voogt (2017) The Evolution of Cubic Dice: From the Roman Through Post-Medieval Period in the Netherlands. Acta Archaeologica, Vol. 88, No. 1, pp. 163–173.

Etchells, Pete (2019) Lost in a Good Game: Why We Play Video Games and What They Can Do for Us. London: Icon Books Ltd.

Finkel, Irving (2007) On the Royal Game of Ur. In Irving Finkel (ed.) Ancient Board Games in Perspective. London: British Museum Press, pp. 16–32.

Foster, Jonathan (2013) The Story of Cluedo. York: Inventive Publishing.

Gasser, R. (1996) Solving Nine Men's Morris. Computational Intelligence, Vol. 12, pp. 24–41.

Gerdes, Paulus (1985) Three Alternate Methods of Obtaining the Ancient Egyptian Formula for the Area of a Circle. Historia Mathematica, Vol. 12, No. 3, pp. 261–268.

Hall, Wm. B., Jr. (1895) The Evolution of Whist. Sewanee Review, Vol. 3, No. 4, pp. 457–467.

Hansen, D. M. & Hansen, K. D. (2021) Clues About Bluffing in Clue: Is Conventional Wisdom Wise? IEEE Transactions on Games, Vol. 13, No. 3, pp. 310–314.

Hesse, Hermann (1943) The Glass Bead Game. New York: Holt, Rinehart and Winston.

Horadam, F. (1978) Wythoff Pairs. Fibonacci Quarterly, Vol. 16, pp. 147–151.

Horn, F. & de Voogt, A. (2009) The Development and Dispersal of l'Attaque Games. In Jorge Nuno Silva (ed.) Proceedings of Board Game Studies Colloquium XI. Lisbon: Associação Ludus, pp. 43–52.

Huizinga, Johan (1938) Homo Ludens. New York: Random House.

Irving, Geoffrey & Donkers, Jeroen & Uiterwijk, Jos (2003) Solving Kalah. ICGA Journal, Vol. 23, No. 3, pp. 139–147.

John, Angela V. & Guest, Revel (1989) Lady Charlotte: A Biography of the Nineteenth Century. London: Weidenfeld & Nicolson.

Jung, W. H. & Kim, S. N. & Lee, T. Y. & Jang, J. H. & Choi, C. H. & Kang, D. H. & Kwon, J. S. (2013) Exploring the Brains of Baduk (Go) Experts: Gray Matter Morphometry, Resting-State Functional Connectivity, and Graph Theoretical Analysis. Frontiers in Human Neuroscience, Vol. 7, 663, pp. 1–17.

Kaplan, G. (2020) Play Behaviour, Not Tool Using, Relates to Brain Mass in a Sample of Birds. Scientific Reports, Vol. 10, 20437, pp. 1–15.

Kawabata, Yasunari (1972) The Master of Go. New York: Alfred A. Knopf. Originally published in Tokyo, Japan, by Shincho magazine in 1951.

Kraaijeveld, A. R. (2001) Origin of Chess—a Phylogenetic Perspective. Board Games Studies, Vol. 3, pp. 39–50.

Kyriakidis, E. (2017) Rituals, Games and Learning. In C. Renfrew, I. Morley & M. Boyd (eds.) Ritual, Play and Belief, in Evolution and Early Human Societies. Cambridge: Cambridge University Press, pp. 302–308.

Lam, C. W. H. (1991) The Search for a Finite Projective Plane of Order 10. American Mathematical Monthly, Vol. 98, No. 4, pp. 305–318.

Lampis, M. & Mitsou, V. & Sołtys, K. (2012) Scrabble Is PSPACE-Complete. In E. Kranakis & D. Krizanc & F. Luccio (eds.) Fun with Algorithms. FUN 2012. Lecture Notes in Computer Science, Vol. 7288. Berlin: Springer.

Lecoq, Jacques (2000) The Moving Body: Teaching Creative Theatre. London: Methuen.

Lo, Andrew (2000) The Game of Leaves: An Inquiry into the Origin of Chinese Playing Cards. Bulletin of the School of Oriental and African Studies, University of London, Vol. 63, No. 3, pp. 389–406.

McMahon, Liz & Gordon, Gary & Gordon, Hannah & Gordon, Rebecca (2017) The Joy of SET: The Many Mathematical Dimensions of a Seemingly Simple Card Game. Princeton, NJ: Princeton University Press.

Michie, Donald (1963) Experiments on the Mechanization of Game-Learning Part I. Characterization of the Model and Its Parameters. Computer Journal, Vol. 6, No. 3, pp. 232–236.

Mkondiwa, M. (2020) Mancala Board Games and Origins of Entrepreneurship in Africa. PLoS ONE, Vol. 15, No. 10, pp. 1–23.

Morewedge, Carey K. & Krishnamurti, Tamar & Ariely, Dan (2014) Focused on Fairness: Alcohol Intoxication Increases the Costly Rejection of Inequitable Rewards. Journal of Experimental Social Psychology, Vol. 50, pp. 15–20.

Murray, H. J. R. (1913) A History of Chess. Oxford: Clarendon Press.

Murray, H. J. R. (1952) A History of Board-Games Other than Chess. Oxford: Oxford University Press.

Neller, Todd & Markov, Zdravko & Russell, Ingrid (2006) Clue Deduction: Professor Plum Teaches Logic. In Proceedings of the Nineteenth International Florida Artificial Intelligence Research Society Conference, Melbourne Beach, Florida, May 11–13, pp. 214–220.

Neufeld, E. (2002) Clue as a Testbed for Automated Theorem Proving. In R. Cohen & B. Spencer (eds.) Advances in Artificial Intelligence. Canadian AI 2002. Lecture Notes in Computer Science, Vol. 2338. Berlin: Springer.

Osborne, Jason A. (2003) Markov Chains for the RISK Board Game Revisited. Mathematics Magazine, Vol. 76, No. 2, pp. 129–135.

Parlett, David (1990) The Oxford Guide to Card Games. Oxford: Oxford University Press.

Parlett, David (1999) The Oxford History of Board Games. Oxford: Oxford University Press.

Penco, Carlo (2013) Dummett and the Game of Tarot. Teorema, Vol. 32, No. 1, pp. 141–155.

Rendell, P. (2016) Turing Machine in Conway Game of Life. In A. Adamatzky & G. Martínez (eds.) Designing Beauty: The Art of Cellular Automata. Emergence, Complexity and Computation, Vol 20. Cham: Springer.

Richmond, John (1994) The "Ludus Latrunculorum" and "Laus Pisonis" 190–208. Museum Helveticum, Vol. 51, No. 3, pp. 164–179.

Rollefson, Gary O. (1992) A Neolithic Game Board from 'Ain Ghazal. Jordan Bulletin of the American Schools of Oriental Research, No. 286, pp. 1–5.

Rowe, M. W. (1992) The Definition of "Game." Philosophy, Vol. 67, No. 262, pp. 467–479.

Schadd, Maarten & Winands, Mark & Uiterwijk, Jos & Herik, H. & Bergsma, Maurice (2008) Best Play in Fanorona Leads to Draw. New Mathematics and Natural Computation, Vol. 4, No. 3, pp. 369–387.

Schreiber, C. (1901) Catalogue of the Collection of Playing Cards Bequeathed to the British Museum by Lady Charlotte Schreiber. London: Longmans & Co.

Schuster, S. (2017) A New Solution Concept for the Ultimatum Game Leading to the Golden Ratio. Scientific Reports, Vol. 7, 5642, pp. 1–11.

Silber, R. (1976) A Fibonacci Property of Wythoff Pairs. Fibonacci Quarterly, Vol. 14, pp. 380–384.

Silber, R. (1977) Wythoffs Nim and Fibonacci Representations. Fibonacci Quarterly, Vol. 15, pp. 85–88.

Solnick, Sara J. (2001) Gender Differences in the Ultimatum Game. Economic Inquiry, Vol. 39, pp. 189–200.

Solnick, Sara J. & Schweitzer, Maurice E. (1999) The Influence of Physical Attractiveness and Gender on Ultimatum Game Decisions. Organizational Behavior and Human Decision Processes, Vol. 79, No. 3, pp. 199–215.

Stewart, Ian (1996) How Fair Is Monopoly? Scientific American, Vol. 274, No. 4, pp. 104–105.

Stewart, Ian (1996) Monopoly Revisited. Scientific American, Vol. 275, No. 4, pp. 116–119.

Suits, Bernard (1978) The Grasshopper: Games, Life and Utopia. Toronto: University of Toronto Press.

Sykes, M. G. (1984) Bingo! Mathematical Gazette, Vol. 68, No. 444, pp. 98–103.

Tan, Bariş (1997) Markov Chains and the RISK Board Game. Mathematics Magazine, Vol. 70, No. 5, pp. 349–357.

Thi Nguyen, C. (2020) Games: Agency as Art. Oxford: Oxford University Press.

Topsfield, Andrew (2006) Snakes and Ladders in India: Some Further Discoveries. Artibus Asiae, Vol. 66, No. 1, pp. 143–179.

Van Ditmarsch, H. (2001) Knowledge Games. Bulletin of Economic Research, Vol. 53, pp. 249–273.

Verne, Jules (1873) Around the World in Eighty Days. Paris: Pierre-Jules Hetzel.

Von Leyden, Rudolf (1982) Ganjifa. The Playing Cards of India. London: Victoria and Albert Museum.

Wardhaugh, Benjamin (2012) Poor Robin's Prophecies. Oxford: Oxford University Press.

Witter, R. Teal & Lyford, Alex (2020) Applications of Graph Theory and Probability in the Board Game Ticket to Ride. In International Conference on the Foundations of Digital Games (FDG '20). New York: Association for Computing Machinery, Article 65, pp. 1–4.

Wittgenstein, Ludwig (1953) Philosophical Investigations. Oxford, UK: Basil Blackwell.

Yan, Jeff & Randell, Brian (2005) A Systematic Classification of Cheating in Online Games. In Proceedings of 4th ACM SIGCOMM Workshop on Network and System Support for Games (NetGames '05). New York: Association for Computing Machinery, pp. 1–9.

INDEX

INDEX

Oxford University Images/Joby Sessions

MARCUS DU SAUTOY is Simonyi Professor for the Public Understanding of Science and professor of mathematics at the University of Oxford. He is author of eight books and two plays and has made numerous programs with the BBC. Du Sautoy is a Fellow of the Royal Society and recipient of many awards, including the Berwick Prize and an OBE. He lives in London.